GETTING TO KNOW
ArcGIS® Pro 3.2

GETTING TO KNOW

Arc**GIS**®

PRO 3.2

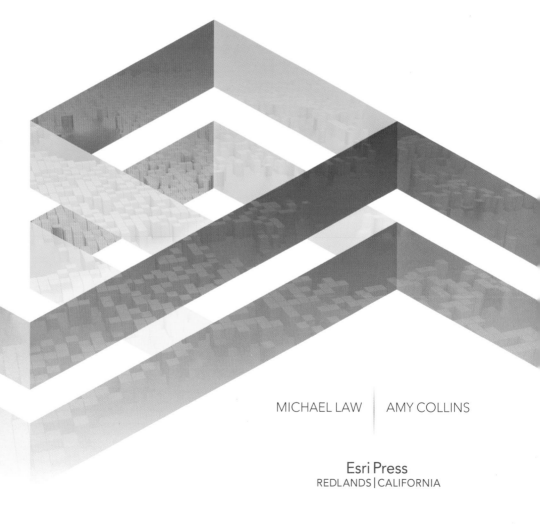

MICHAEL LAW | AMY COLLINS

Esri Press
REDLANDS | CALIFORNIA

Esri Press, 380 New York Street, Redlands, California 92373-8100
Copyright © 2024 Esri
Fifth edition
All rights reserved.

Printed in the United States of America
28 27 26 25 24   1 2 3 4 5 6 7 8 9 10

ISBN: 9781589487772
Library of Congress Control Number: 2023951938

Contents

Preface vii

Chapter 1   Introducing GIS 1
Exercise 1   Explore ArcGIS Online. 11

Chapter 2   A first look at ArcGIS Pro 33
Exercise 2a   Learn some basics 34
Exercise 2b   Go beyond the basics 49
Exercise 2c   Experience 3D GIS. 59

Chapter 3   Exploring geospatial relationships 69
Exercise 3a   Extract part of a dataset 71
Exercise 3b   Incorporate tabular data 79
Exercise 3c   Calculate data statistics 95
Exercise 3d   Connect spatial datasets 104

Chapter 4   Creating and editing spatial data 113
Exercise 4a   Build a geodatabase 115
Exercise 4b   Create features 126
Exercise 4c   Modify features 138

Chapter 5   Facilitating workflows 153
Exercise 5a   Manage a repeatable workflow using tasks. 156
Exercise 5b   Create a geoprocessing model 169
Exercise 5c   Run a Python command and script tool 182

Chapter 6   Geoenabling your project **195**

Exercise 6a   Prepare project data. 198

Exercise 6b   Geocode location data 204

Exercise 6c   Use geoprocessing tools to analyze vector data. 217

Chapter 7   Analyzing spatial and temporal patterns **229**

Exercise 7a   Create a kernel density map 231

Exercise 7b   Perform a hot spot analysis 235

Exercise 7c   Explore the results in 3D 244

Exercise 7d   Animate the data. 251

Chapter 8   Determining suitability. **257**

Exercise 8a   Prepare project data. 261

Exercise 8b   Derive new surfaces 268

Exercise 8c   Create a weighted suitability model. 278

Chapter 9   Presenting your project **287**

Exercise 9a   Apply detailed symbology 290

Exercise 9b   Label features. 299

Exercise 9c   Create a page layout. 306

Exercise 9d   Share your project 317

Glossary 321

Task index 329

Image and data source credits 333

Preface

The title sums it up: this book is for those who want to get to know ArcGIS® Pro—a new generation of geographic information system (GIS) software from Esri®. Whether you are a student in an introductory GIS course, an at-home learner who wants to build a foundational knowledge of GIS, or a professional who is considering adding GIS to your resources, this book is for you. No prior GIS software knowledge is required or assumed. This book is also suitable for those who are used to a different GIS product and want to see how to do familiar tasks in a new environment.

The primary focus is, naturally, ArcGIS Pro, but because of the integrated design of the ArcGIS platform, other ArcGIS components are incorporated as well, such as ArcGIS Online.

A word about scope—although this workbook is designed to provide a broad overview of ArcGIS Pro, a truly comprehensive manual would be massive. Instead, we aim to provide a diverse sampling of industries, scenarios, and workflows that highlight the broad appeal and many core functions offered by GIS and ArcGIS Pro. At the same time, we try to keep the book's length reasonable—something that a student in a classroom can feasibly complete in a quarter or semester. When you complete this book, you should feel comfortable enough with ArcGIS Pro to start working with it on your own.

About the fifth edition

This edition has been updated throughout with screen captures and instructions to work with the updated software.

Book features

The book has nine chapters, each containing the following features, which are designed to facilitate an efficient and effective learning process.

Exercise objectives

Exercises are composed of learning objectives, which are listed for each chapter and repeated as headings throughout the chapter exercises. Each objective is accomplished by following a sequence of steps. Using objective headings helps break each exercise into logical chunks and provides a reminder of why you are clicking this or that button, option, or command.

Data list

In the real world, you do not begin a geospatial analysis project before first gathering relevant data. Therefore, we list the student data, with a brief description of what it is and where it comes from, at the start of each chapter.

Exercise workflow

As an expansion of the exercise objectives, each exercise begins with a summary of the workflow, explaining the "what, why, and how" of the upcoming exercise. This description will help you understand the bigger picture rather than get muddled in a sea of instructions.

A note about exercise scenarios—many of them are based on real-world projects; however, the data and workflows are usually simplified for training purposes. These exercises are meant to teach software and data management skills in a realistic setting; they are not meant to be an authoritative guide to geographic problem solving.

GIS in the world

These short sidebars highlight real-world GIS problem solving and offer a link to read more.

Tips and questions

Reminders, shortcuts, or alternative approaches are sprinkled throughout each chapter. Questions keep learners actively involved. Answers can be found in the book's online resources at links.esri.com/GTKPro3.2.

Summary

The summary offers a brief recap of what you have learned in each chapter.

Glossary terms

Shown in colored text, glossary terms are listed at the end of each chapter and defined in the glossary at the end of the book.

Hardware and software requirements

To perform the exercises in this book, you need ArcGIS Pro installed on a computer that is running the Windows operating system, an internet connection, and a web browser to access ArcGIS Online.

Licensing the software

Use an existing license

If you have existing credentials (or can obtain credentials from your educational institution or organization) that provide access to the required elements of ArcGIS, you may use those credentials and proceed.

Use an evaluation code

This book comes with an evaluation code that will grant you a fully functional, 180-day license. The code can be found inside the back cover of the print book. Codes for e-books are available for purchases made through the delivery platform VitalSource and are redeemable after purchase. Activate your code and license your software at links.esri.com/EVAcode.

Use a trial version

Additional trial account options can be found at links.esri.com/ArcGISTrial.

Installing the exercise data

The exercise data for this book is available for download from ArcGIS Online, available at links .esri.com/GTKPro3.2Data. Click the Getting to Know ArcGIS Pro 3.2 – Exercise Data item to download it. Unzip the file and move it to your C drive. If you are in class, your instructor may provide alternative directions for downloading the data. Exercise data that accompanies this book is covered by a license agreement that stipulates the terms of use.

> *Because many of the exercises require users to modify the original data, we recommend that you make a copy of each chapter's data folder before you start any exercises.*

How to use this book

Each chapter focuses on a unique project and has its own dataset, so theoretically you can do the chapters in any order. But the book is designed for linear progression—that is, chapter 2 has more explanation and more explicit instruction than chapter 9, in which we assume a more GIS-savvy audience. Also, exercises within chapters typically build on each other, so it is advisable to do all the exercises in a chapter in order. If you cannot complete an exercise successfully, most chapters provide interim data (in a Results folder) so that you can continue with the remaining exercises. For more information about other resources, visit links.esri.com/GTKPro3.2.

Acknowledgments

It takes a village to make a book. We are indebted to many individuals at Esri for contributing to the process. Thanks to everyone at Esri Press. And thank you to all the reviewers and testers of this book. Also, to all the individuals and organizations who provided data, graphics, project scenarios, and advice: thank you. This book would not be what it is without your assistance and generosity. A complete list of data contributors can be found at the back of the book.

And thanks to the GIS learners who purchase this book. We hope you enjoy the fifth edition of *Getting to Know ArcGIS Pro*.

Michael Law
Toronto, Ontario, Canada

Amy Collins
Napa, California

## CHAPTER 1

Introducing GIS

Exercise objectives

1: Explore ArcGIS® Online
- Explore a public map.
- Configure the map symbology.
- Configure map pop-up windows.
- Save a map.

What is a GIS?

Probably the most common question to those working in the *GIS* field is also one of the most difficult to answer in just a few brief paragraphs: What is a GIS? A GIS is composed of five interacting parts that include hardware, software, data, procedures, and people. You are likely already familiar with the hardware—computers, smartphones, and tablets. The software consists of both desktop and mobile applications that help make maps. The data is information in the form of points, lines, and polygons that you see on a *map*. People, users like you, learn how to collect data using mobile devices and then make maps using the software and data on computers. As your knowledge of GIS grows, you will learn more about procedures and workflows to make maps for yourself or your organization. Decision-makers and others in an organization rely on GIS staff to maintain data and create insightful map products.

GIS has many facets. It captures, stores, and manages data. It allows you to visualize, question, analyze, and interpret the data to understand relationships, patterns, and trends. GIS can be used simply for mapping and cartography. You can use it on the web to view maps and collections of data. You can also use it to perform spatial analysis to derive information from multiple data sources. In any capacity, the results from a GIS can influence decisions. Organizations in almost every industry, no matter what size, benefit from GIS and realize its value.

Collecting spatial data—that is, information that represents real-world locations and the shapes of geographic features and the relationships between them—involves using coordinates and a suitable map projection to reference this data to the earth. For example, the distance that separates a conservation area and a neighborhood of a city is an example of a spatial relationship. How is wildlife in the conservation area affected by the increasing pressures of a growing urban setting? The spatial relationship between geographic features allows the comparison of different types of data.

When paired with attribute data—information about spatial data—a GIS becomes a powerful tool. For example, the location of a hospital is considered the spatial data (its property boundaries referenced to the earth). Information about the hospital—such as its name, number of rooms available, emergency rooms, specialization in medical procedures, patient capacity, and number of staff—is all considered attribute data. You can use this attribute information to maintain records in a hospital network. It allows people who have this information to perform spatial analysis, a technique that reveals patterns and trends, to answer the following types of questions: What are the average wait times for emergency visits to the hospital? Does the patient capacity efficiently serve the demographics of a given city area? Do certain medical conditions happen more frequently in the area, and is the hospital equipped to handle them? To answer these questions fully, you must compare the data and attempt to explain the patterns. A children's hospital can integrate spatial analysis with population-based resource planning to propose children's health-care initiatives. This integration can greatly increase the hospital's ability to identify and allocate resources to better meet local health-care needs, providing timely access to care for children across a city or region.

GIS today

GIS has been helping people better understand their world since the 1960s. It provides a framework of practical means for transforming the world with all kinds of activities, from improving emergency response to understanding bird migration patterns. People are integrating GIS into how they work with data because it is a visual, quantitative, and analytic tool. It provides people with the structures and concepts to handle data systematically.

People today have unprecedented access to data and information. A growing system of connected networks allows people to easily access data, collaborate with others, and produce and share results from desktops, laptops, and mobile devices—essentially from anywhere. The current trend of connecting people who work in the office and in the field allows for real-time analysis. Decision-makers use operations dashboards to monitor real-time data feeds and other sources of information. For example, a GIS coordinator for a local government can track real-time emergencies and respond by coordinating fire, police, and ambulance resources.

GIS is pervasive, interactive, and social. Dynamic and interactive maps on the internet, known simply as *web maps*, are ideal for allowing many users to access and quickly locate features and visualize data. In the past, it took a team of GIS professionals to put together an online map. Now anyone can connect to ArcGIS Online, make a map with a few layers and a

GIS IN ACTION: HELPING FIGHT HUNGER

The COVID-19 pandemic has added to the challenges faced by Feeding America, a national network of food banks that provide meals to 46 million people each year. The pandemic created a threat to safety and health for which it is estimated that, overall, one in nine people may face hunger. Many families face insecurity not only in food but also in jobs and housing. A local organization, Feeding America Riverside San Bernardino, has seen a 60 percent increase in food needs across the community since the pandemic started. With a need to scale its home delivery in response to requests by families, the Homebound Emergency Relief Outreach (HERO) program was created. A web-based dashboard was created in the Esri mapping environment to allow volunteers to easily see whether a home delivery was near their location, assign themselves to make it, and get on the road. Read more about the project, in *ArcUser*, "A Growing Hunger": www.esri.com/about/newsroom/arcuser/feedingamerica.

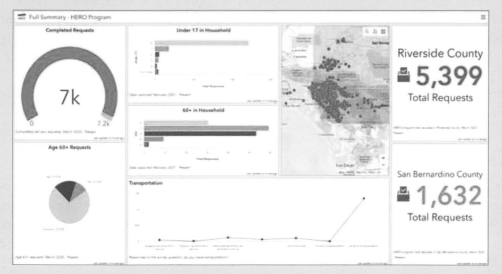

Figure 1.1. The Feeding America HERO program dashboard. Photo courtesy of Feeding America.

basemap, and then share it with friends, coworkers, or anyone. The latest generation of web maps has simplified that process and now forms a platform that anyone can use.

Governments are opening access to data at an unprecedented rate. The *open data* movement provides agencies and the public with authoritative data and enables all levels of government to develop new tools and applications. Typically, only highly sensitive data is safeguarded or copyrighted anymore. Open data provides a way for people to extract information when they need it. It allows citizens, organizations, and governments to get right to problem solving, rather than spending time searching for and requesting data. ArcGIS® Hub—an easy-to-configure community engagement environment that organizes people, data, and tools through information-driven initiatives—is one such ArcGIS Online solution. It allows an organization to host the data it collects so that the public can freely view interactive maps and search for and download data.

Figure 1.2. Governments and organizations share open data through online portals. Michigan Open Data Portal (https://data.michigan.gov).

Basic GIS principles and concepts

You can visualize data in a GIS as layers in a map. You can represent geographic and manu-factured objects on the earth in a map by using symbols: points, lines, and polygons. In the accompanying map of Abu Dhabi, points represent trees and points of interest; lines represent roadways; and the polygons represent building footprints, green space, and water. Point, line, and polygon data is also called *vector* data. Features of the same type—such as trees, roadways, or buildings—are grouped together and displayed as *layers* on a map. To make a map, you add as many layers as you need to tell a story. If you are telling a story about a river that seasonally

GIS IN THE WORLD: WILDFIRE RESPONSE USING DRONE IMAGERY

Devastating wildfires in Northern California have burned hundreds of thousands of acres, destroying thousands of structures and endangering countless homes and lives. Law enforcement in San Mateo and Santa Cruz Counties needed a better way to view areas affected by wildfires and determined that drones or unmanned aerial systems (UAS) were an important part of disaster response. The goal was to define, capture, and process imagery and then share the geospatial data with various emergency response teams using ArcGIS Online. Read more about the project in *ArcNews*, "Rapidly Processed Drone Imagery Improves California Wildfire Response": www.esri.com/about/newsroom/arcnews/rapidly-processed-drone-imagery-improves-california-wildfire-response.

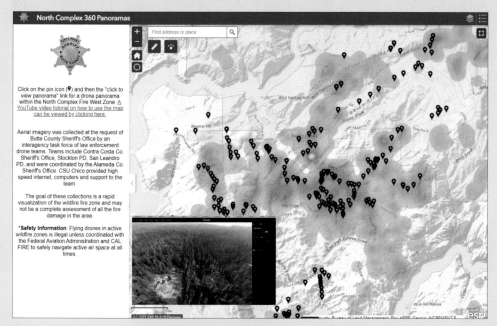

Figure 1.3. Fire incidents within the North Complex Fire West Zone. Courtesy of GeoAcuity.

floods, you add a river layer and past flood hazard layers. You can also add a land-use layer to visualize what type of property, such as agricultural or residential, is affected by the flooded river. If you are building a city map, you start with a boundary layer, a street layer, and building footprints. By adding more layers, you can build a map that describes the city to your readers.

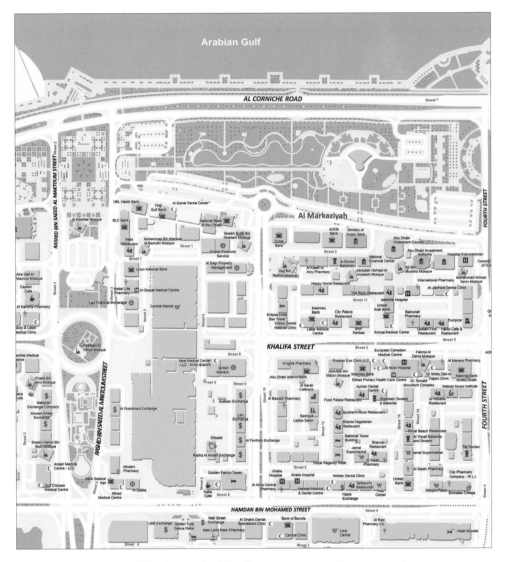

Figure 1.4. Map of Abu Dhabi showing building footprints, points of interest, roads, green space, and water features. Map courtesy of Municipality of Abu Dhabi City.

If you make a map of your house, a lake, or a city park, you might draw an outline to represent the outer boundary. But what about natural phenomena—such as temperature, elevation, precipitation, ocean currents, and wind speed—that have no real boundaries? Weather maps show blue areas for cold and red areas for hot. Wind speed can be represented using a range of colors. Or you can instead record and collect measured values for any location on the earth's surface to form a digital surface, also known as a *raster*. Captured location data is recorded in a matrix of identically sized square cells at a specific resolution—for example, 15 square meters. In the accompanying example, an analysis of an aquifer uses different rasters to calculate a result showing saturated thickness and usable lifetime.

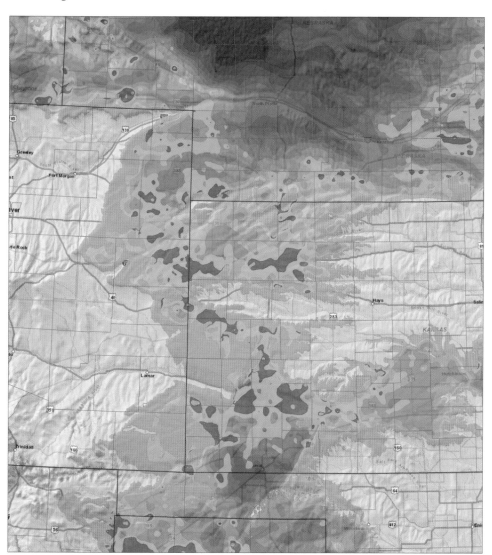

Figure 1.5. Map of the Ogallala Aquifer showing a surface that represents the saturated thickness, water-level change, and projected usable lifetime of the aquifer. Map courtesy of Center for Geospatial Technology.

Features have locational data behind them. Features also contain attribute data, known as *attributes*. For a forestry map, point features that represent trees might include attributes such as tree species, height, bark thickness, and trunk diameter. For a utility map, lines that represent sewer pipes might include attributes such as flow rate, flow direction, pipe material, and length. Feature attribute information is stored in a table in a GIS database. Each feature occupies a row in the table, and an attribute field occupies a column. A GIS database can hold large collections of features and their corresponding attribute data. A GIS provides many tools for you to query, manipulate, and summarize large quantities of data.

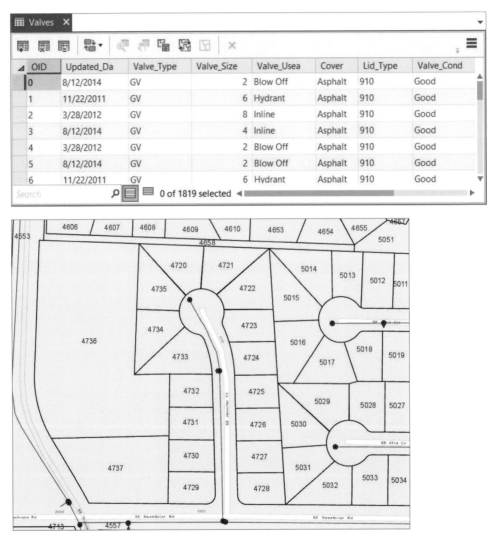

Figure 1.6. A map with parcels, water lines, and valves. Attributes of the valve features include maintenance dates and information about valve type, size, usage, cover, lid type, and condition.

Data can be queried and analyzed. In a GIS, you can perform a query on all the data that relates to phrases, terms, or features that you choose. For example, you might be looking for clusters of low-income neighborhoods to analyze poverty levels per square mile. Querying data from a database allows you to display only the data that relates to a certain theme. Additionally, a GIS enables you to identify spatial patterns in the data using geospatial processing tools. What is the problem you are trying to solve and where is it located? The accompanying map shows analysis and a complex pattern of senior citizen out-migration. Depending on your project, you can choose from among hundreds of analysis tools.

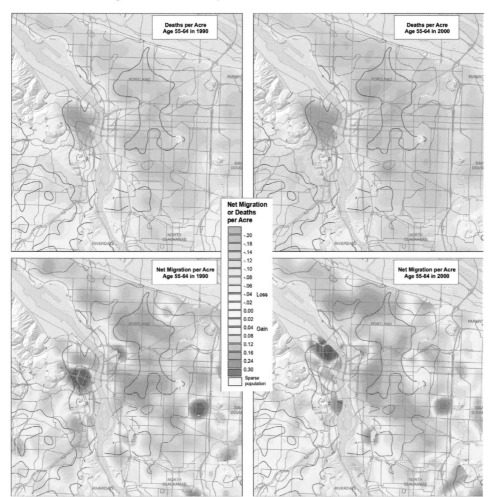

Figure 1.7. Map of Portland, Oregon, showing net migration or deaths per acre because of "senior shedding" or out-migration. The red isolines identify concentrated areas in which mothers age 30 and over gave birth. Map courtesy of Portland State University.

ArcGIS

ArcGIS Pro is designed for GIS professionals to analyze, visualize, edit, and share maps in 2D and 3D.

ArcGIS Desktop is part of the much larger ArcGIS suite of software, which also includes ArcGIS Online and ArcGIS Enterprise. Organizations can use the entire suite to share maps and apps with their users.

ArcGIS includes ready-to-use spatial data and related GIS services, such as global basemaps, high-resolution imagery, demographic reports, lifestyle data, geocoding and routing, hosting, and more.

Finally, ArcGIS includes essential tools for developers to build web, mobile, and desktop apps.

ArcGIS Pro

In this book, you will learn how to use ArcGIS Pro and ArcGIS Online. Your work in ArcGIS Pro is organized into *projects*. These projects contain maps, layouts, layers, tables, tasks, tools, and connections to servers, databases, folders, and styles. Essentially, all the resources needed for a project are in one place—in a project. ArcGIS Pro can also connect to ArcGIS Online public content. And if you belong to an *organization*, you can share the content with your team. Projects are designed to be collaborative so that others can share and open them.

Maps and layouts display a project's spatial data in either 2D maps or 3D scenes, or both simultaneously. You can create, view, and edit multiple maps, layouts, and scenes side by side, and even link them so that they can be panned and zoomed together. ArcGIS Pro uses ArcGIS Online basemaps, which provide a backdrop or frame of reference as you add your own layers.

A collection of *geoprocessing* tools allows you to perform spatial analysis and manage GIS data. Geoprocessing involves an operation that manipulates spatial data, such as creating a new dataset or adding a field to a table. You can combine tools in ModelBuilder™ to create a diagram or model of your spatial analysis or data management process. For advanced users, Python, the scripting language of ArcGIS, provides a way to write custom scripting functions to help automate ArcGIS workflows. In addition, tasks can be created and defined for organizational users who are required to follow specific workflow steps.

The ability to share your work is a central part of ArcGIS. In ArcGIS Pro, you can share maps, layers, or entire projects. Sharing involves packaging components into a compressed file, which you can distribute to others within your organization or externally. You can store your package on a shared network drive or serve it across a website or mobile device.

Exercise 1: Explore ArcGIS Online

Estimated time to complete: 30 minutes

You will begin your ArcGIS journey by signing in to an ArcGIS Online organizational account and exploring a public map of traffic incidents within walking areas of public schools in Washington, DC. The goal is to identify areas where students may be at risk during travel to and from school based on driving incidents that have occurred. You will then configure a new map to better visualize the data and familiarize yourself with Map Viewer in ArcGIS Online.

Exercise workflow

- Sign in to ArcGIS Online and explore a public map that shows public schools and school walking time areas.
- Configure symbology for public schools and school walking areas.
- Configure map pop-up windows for readability.
- Configure traffic safety incidents and apply filters and clustering.
- Save the map to your My Content page.

Because of the dynamic nature of websites, the appearance or options of ArcGIS Online may change at any time.

Explore a public map

Some exercises in this book require you to sign in to your ArcGIS Online organizational account. An organizational account provides a suite of ready-to-use apps that run on browsers, desktops, and mobile devices; access to maps and data; ArcGIS Online service credits; and more.

You will visit ArcGIS Online to explore traffic incident data around public schools in Washington, DC.

1. Go to www.arcgis.com.

2. Sign in using your ArcGIS Online account credentials.

   You are redirected to your organization's main page.

Your profile contains your user settings. It allows you to store your own content. For more information on how to manage your profile (https://doc.arcgis.com /en/arcgis-online/get-started/profile.htm), consult ArcGIS Online Help at https://doc.arcgis.com/en/arcgis-online and browse to Get Started > Set Up Account > Manage Profile and Settings.

3. On the upper right, click the magnifying glass to open the search box, type **owner: GettingToKnowArcGISPro3.2 DC Public School Safety Map**, and press Enter.

Adding this text (owner:GettingToKnowArcGISPro3.2) limits the search results to content uploaded by the Esri Press organizational account associated with this book. Your search results may not return anything. By default, search results are limited to content within your organization, so you will set an option to show results from all public content in ArcGIS Online.

4. Under Content, turn off the option to Only Search in Your ArcGIS Organization.

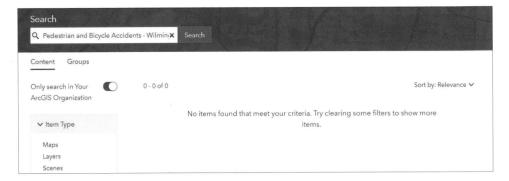

A web map named *DC Public School Safety Map* appears in the search results.

5. Click the title to see its details pane. On the details pane, if necessary, click View Item Details.

The map description provides an overview about the map, including the owner, description, map layer contents, and other properties.

When they click View Item Details, users have the option to access both Map Viewer Classic and Map Viewer in ArcGIS Online. This chapter uses the latest version of Map Viewer.

6. Click Open in Map Viewer.

Depending on your organizational settings, the button may say Open in Map Viewer Classic, in which case you should click the down arrow and click Open in Map Viewer. The map shows public schools in Washington, DC, over the default topographic basemap.

7. On the Contents toolbar (the left vertical menu), click Expand at the bottom left, if necessary, and click Layers.

The Layers pane contains three layers—a layer for public schools, a layer of school service areas by walking times, and a layer of the Vision Zero Safety incidents collected by the City of Washington, DC. Layer visibility can be turned on and off by clicking the Visibility button (eye icon). One or more of these layers may be turned off when you open Map Viewer.

You will change the basemap to see the precise locations of accidents. Basemaps provide a backdrop and frame of reference for operational layers, such as accidents. As you zoom in, the basemap provides more detail. In this case, you see topographic details. In other basemaps, you might see the ocean floor. In the imagery basemap, for example, you can see your house, school, or workplace.

The vertical menu on the left is the Contents toolbar. The vertical menu on the right is the Settings toolbar.

8. On the Contents toolbar, click Basemap to show all available basemaps. Click the Streets basemap, close the Basemap pane, and zoom or pan around until your map extent roughly matches the figure.

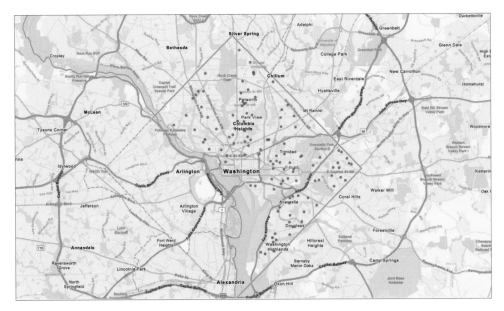

9. Click Layers to see the Layers pane again. Drag the Public Schools layer to the top of the Layers pane.

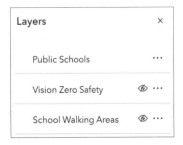

For each layer, you can see the Visibility button and Options. If you click the Options button, more actions will be revealed on the drop-down menu that allow you to zoom to the layer, rename the layer, show the attribute table, show layer properties, or remove the layer. You can click the Options button again to hide it.

When you click any layer, an additional context pane on the right of the window appears. This context pane allows you to view and adjust layer properties, *styles*, filters, pop-ups, labels, and more. If you can't see the full context pane, click Expand at the bottom right of the Settings toolbar. Clicking any of the context pane menu items affects only the current selected layer. Clicking the same Settings toolbar item will collapse its options.

Feel free to explore the map and turn on the visibility of the other layers.

10. Make Public Schools the only visible layer.

Next, you will change the layer symbology to something more meaningful.

Configure the map symbology

1. Click the Public Schools layer to select it. If the Properties pane isn't visible, click Properties on the Settings toolbar at the far right to view the current symbology.

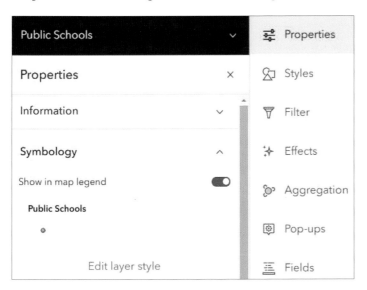

2. Under Symbology, click Edit Layer Style.

The Styles pane shows that the layer is symbolized based on Location (single symbol) to display the public schools. Next, you will change the symbol to something different to help differentiate it.

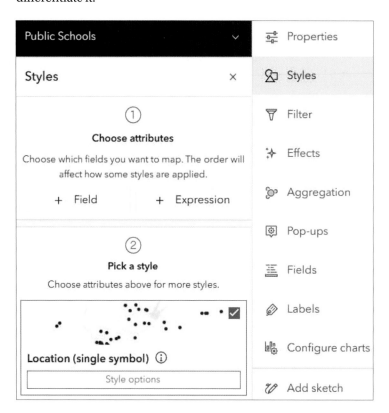

3. Click Style options.

4. In the Style Options pane, click the Symbol style to change the public schools symbol.

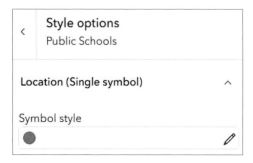

5. In the Symbol Style pane that appears, under Current Symbol, click the right arrow inside the box to view the Change Symbol settings. Under Category, click the down arrow, and choose POI under Vector Symbols.

6. In the icon palette, scroll down and click the purple School symbol and click Done. (Point to each symbol in the icon palette to see its pop-up label.) In the Symbol Style pane, adjust the symbol size to **20**.

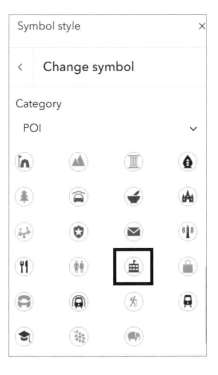

7. At the bottom of the Style Options pane, click Done to apply the symbol adjustments and close that pane. Click Done again to close the Styles pane.

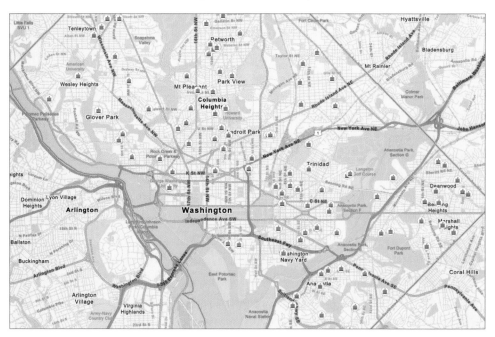

8. In the Layers pane, for the School Walking Areas layer, click the Visibility button to make the layer visible on the map. Click the School Walking Areas layer name in the Layers pane to activate it.

Activating a layer in the Layers pane allows you to modify styles and other settings.

9. If necessary, reopen the Styles pane and make sure that School Walking Areas is the active layer (shown across the top of the pane). If the Public Schools layer is still the active layer, click the down arrow at the top, and choose School Walking Areas.

10. Under Choose Attributes, click the Add (+) Field button to add a field. In the Add Fields list, click ToBreak, and click Add.

The ToBreak attribute shows how many minutes it takes to walk to the school.

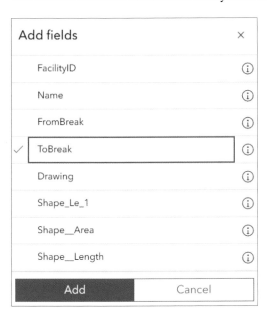

11. Under Pick a Style, scroll down and click Types (unique symbols).

12. Click Style Options to edit its options.

13. Under Symbol style, click the color ramp. In the Symbol Style pane, under Fill Color, click the color ramp to open the Ramp pane.

14. In the Ramp pane, click Flip Ramp Colors, and click Done to close the Ramp pane.

15. In the Symbol Style pane, under Outline Color, click the No Color button at the far right to turn off the outline.

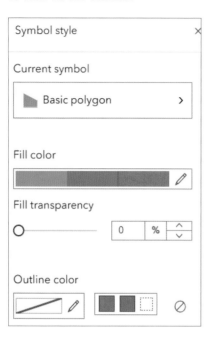

16. At the bottom of the Style Options pane, click Done, and click Done again to apply the symbol adjustments.

17. On the Settings toolbar, click Properties. Under Appearance, adjust the Transparency slider to **50%**.

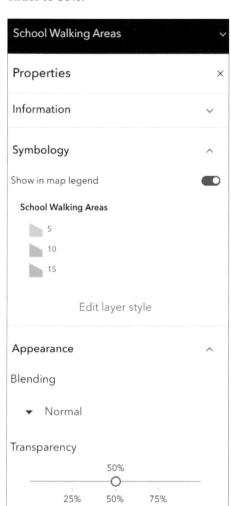

18. Click Properties on the Settings toolbar to collapse the pane. Zoom in to the map and pan around to see more detail.

You can now see school walking areas more clearly with detail from the basemap still visible. The green areas show a five-minute walk to the nearest school, the blue shows a 10-minute walk, and the red shows a 15-minute walk.

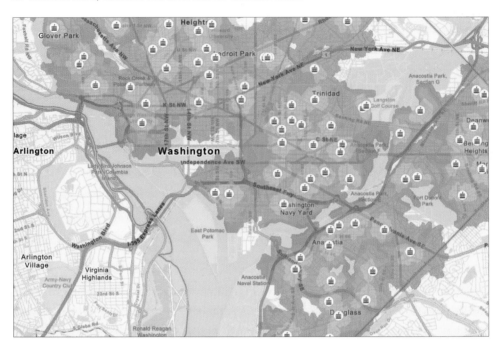

ON YOUR OWN

Experiment with symbolizing the layers to practice and become comfortable with styling. Pay attention to how symbol size and map scale relate to each other and how they appear on the map. There is also an option for symbols to adjust size automatically. Change the color ramps of the school walking areas to see how it affects the visualization of the map based on its classification.

In the following two sections, you will customize this map further by configuring descriptive information about traffic incident features in pop-up windows. You will look at the more serious incidents (such as speeding) that have been officially recorded that occur within these school walking areas. This can give an indication of dangerous areas that may need further intervention by the city to conduct studies on traffic patterns and volume, speed limits, and introducing safety protocols.

Configure map pop-up windows

1. Turn off the visibility for Public Schools and School Walking Areas, and turn on the visibility for Vision Zero Safety. Select Vision Zero Safety in the Layers pane to activate the layer.

2. Zoom in, if necessary, and click an incident location on the map to view its pop-up.

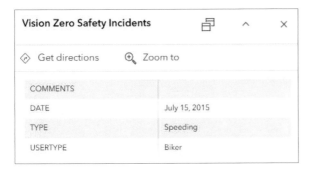

    In the pop-up, you will see four attributes—the date of the incident, the incident type, the user type, and a place for comments. Pop-ups can be configured to show numerous fields and values sourced from the layer's attributes. Although having all this information can be useful, it is unnecessary to show all of it for display purposes. You will adjust which attribute fields are displayed so that you can more easily understand the type of incident that occurred.

3. In the pop-up, click Zoom To to zoom to the single incident, selected on the map in cyan. Click Zoom To again to zoom in further. Using your mouse wheel, zoom out a bit to see a larger grouping of incidents. Close the pop-up.

4. On the Settings toolbar, click Pop-ups to open the Pop-ups pane.

5. In the Pop-ups pane, click Fields List.

    Under Select Fields, you can see the four fields that are currently displayed: COMMENTS, DATE, TYPE, and USERTYPE. Each field has an X beside it for deleting that field.

6. Click Select Fields to see the full list of available fields. Scroll down to USERTYPE, and click the check mark beside it to delete it from your pop-ups. Click Done to return to the Pop-ups pane.

7. If necessary, reopen a pop-up to see the fields that are still displayed.

8. Under Fields List, in the Title text field, type **Incident Types with Comments**.

9. If necessary, reopen a pop-up to see the new title.

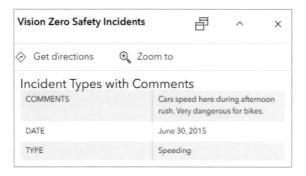

If you accidentally removed a field, you can click Select Fields to view the attribute list where you can add them back to the pop-up list.

10. Click Select Fields, and check USERTYPE to add it back to your pop-up. Click Done.

11. On the Settings toolbar, click Fields.

You can also change the display name of a field to improve readability.

12. In the Fields pane, click USERTYPE to open the Formatting pane. Under Display Name, type **USER**, and click Done.

13. Return to the map, and if necessary, reopen a pop-up to see it fully configured. (Some pop-ups may not have comments added.)

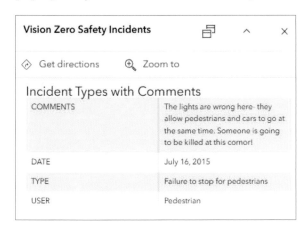

14. Close the pop-up window and the Fields pane.

Next, you will customize this map further by configuring how the Vision Zero Safety traffic incidents look on the map.

Configure filter and clustering

1. Make sure that Vision Zero Safety is the only visible layer. Adjust the zoom level to see the incidents across the city. Make sure the Vision Zero Safety layer is selected (indicated by the blue highlight to the left of its name).

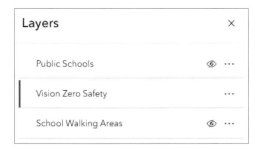

2. On the Settings toolbar, click Filter.

3. In the Filter pane, click Remove Filter to remove existing filter settings.

4. Click Add Expression. Under Expression, click the top field's down arrow, and click TYPE. In the middle field, keep the Is option. In the bottom field, click Speeding.

    This expression filters only the speeding incidents. Other incidents that don't match the filter are unavailable on the map. You'll add a few more filters of more serious incidents.

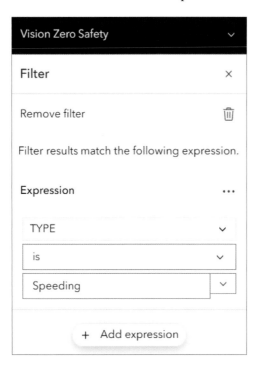

5. Click Add Expression. For the second instance, create an expression where TYPE Is Failure to Stop for Pedestrians.

6. Under Filter results, click the down arrow and choose Match All Expressions (if it is not already chosen).

    Notice that all the incidents on the map become gray (unavailable). This happens because the filter is trying to match both expressions (represented by the And between each expression). You will need to adjust the Filter Results to return to Match at Least One Expression.

7. Under Filter Results, click the down arrow, and choose Match at Least One Expression.

    The incidents that match either speeding or failure to stop for pedestrians now appear on the map, and other incidents are unavailable.

8. Create two more expressions (for a total of four in the set):

   TYPE Is Red Light Running
   TYPE Is Stop Sign Running

9. Click Save at the bottom of the pane to save all the expressions.

   Although the map now shows many points based on the filters you created, enabling clustering will help show where many incidents occur closer together. You will also adjust the point style.

10. On the Settings toolbar, click Styles. Click Style Options to open the Style Options pane.

11. Under Location (Single Symbol), click the symbol in the text field to edit it.

12. In the Symbol Style pane, under Current Symbol, click the symbol to change its category and size.

13. Under Category, use the down arrow to choose Basic Shapes under Classic Symbols. Click the circle in the Basic Shapes set of four shapes, and click Done to return to the Symbol Style pane. Under Size, adjust the size to **4**. Under Fill Color, choose red for fill color (#ff4040).

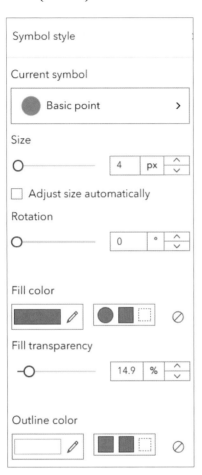

14. At the bottom of the Style Options pane, click Done, and then click Done again.

15. On the Settings toolbar, click Aggregation. Click the Enable Aggregation toggle button to turn it on.

16. Under Clustering, click Options. In the Clustering pane, drag the Cluster Radius slider to the middle (as near as you can place it). Click the blue line segment on the Size Range slider and move it to the far left. Drag the right side of the Size Range slider to the middle.

You see how the clusters change as you adjust the slider. Larger circles represent more occurrences of incidents. Each circle is labeled with a number indicating the number of individual incidents within each cluster.

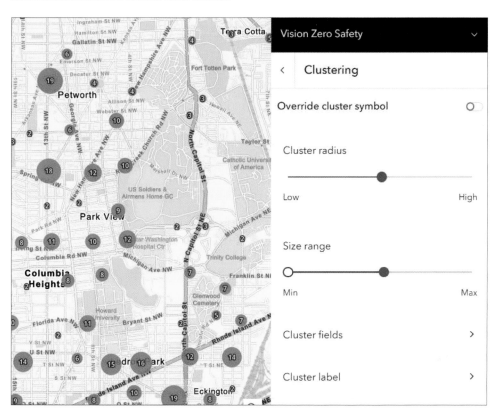

17. Click Aggregation on the Settings toolbar to close the Clustering pane, and click a cluster on the map.

A Cluster Summary pop-up appears, showing how many features that cluster has.

18. In the pop-up, click Browse Features.

Each feature in the cluster is now itemized on its own line.

19. Click any of the incidents in the pop-up to see its unique attributes, and click Zoom To to see that feature on the map.

You can also click Get Directions to map a route from your location and set driving times to help with planning. Doing so adds a Route layer to your Layers pane.

Next, you'll finalize the map and prepare to save it.

Save a map

As an organizational account user, you can save the map to your own workspace. By default, maps that you make in Map Viewer are private (only you can see them). You will save your map and add it to your own content.

You are allowed to save a copy of any map that you work on unless the original map author has enabled Save As protection. You cannot update any existing layers that belong to the original map author. However, you can save any changes you make to the map, which then only you can see.

1. On the Contents toolbar, click Layers. Make all the layers visible.

2. Zoom in and inspect some of the schools and walking areas. Some areas have more incidents than others. Locate a school like one of those located in the figure.

    These areas have larger cluster circles and could pose more risk to students and the community.

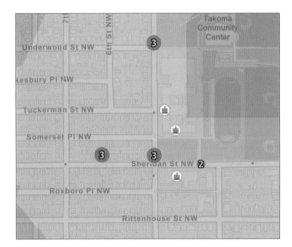

The city may use this information to analyze areas around schools more closely and implement speed limit reductions or introduce more crossing guards at specific intersections.

3. Close the Layers pane, and zoom to a desired extent to see all the layers. Click Save and Open, and click Save As.

4. In the Save Map dialog box, change the title to **DC Public School Safety At Risk**.

5. Add several tags: **accidents**, **safety**, **school safety**. Add a comma after typing a tag. Keep the Folder name.

6. For Summary, type **A map showing safety incidents within walking range of public schools.**

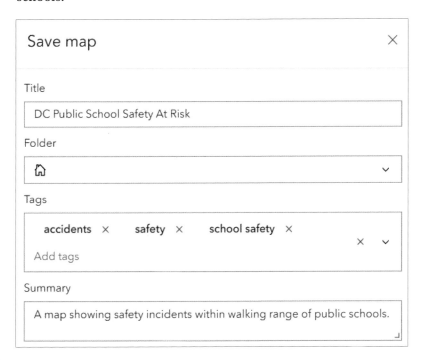

7. Click Save.

You have saved your first ArcGIS Online map. All saved maps appear in My Content and are accessible at any time. Even though the original map was shared with everyone, the map you worked on is considered private and is not shared with anyone in the organization, a group, or the public. You do not need to share this map right now. However, you will learn how to share maps and other content throughout this book because this process is an essential part of ArcGIS.

ON YOUR OWN
Go to My Content and find your own map.

Summary

This chapter introduced you to some background information about what GIS is and how ArcGIS is structured. You started with opening a public map, one that everyone has access to, to see how easy it is to view data and basemaps. You learned how to sign in to an organizational account, search for a map, and open it. You learned how to configure symbology, clusters, and pop-ups to make the data layers and their attributes more usable. You then learned how to save the map to your own organizational account.

You should be getting comfortable using ArcGIS Online. You will use it throughout the book, in which you will learn how to share layers, export content to your organizational account, and open maps in ArcGIS Pro. In chapter 2, you will begin using ArcGIS Pro.

Glossary terms

Glossary terms are listed in the order they appear.

GIS
map
basemap
open data
vector
layer
raster
attribute
ArcGIS Pro
project
organization
geoprocessing
styles

A first look at ArcGIS Pro

Exercise objectives

2a: Learn some basics

- Start a new project.
- Import a map document.
- Create a folder connection.
- Modify map contents.
- Explore the map.
- Examine the contextual ribbon.
- Examine feature attributes.
- Select features.

2b: Go beyond the basics

- Modify feature symbols.
- Label features.
- Measure distances.
- Add a basemap.
- Package and share the map.

2c: Experience 3D GIS

- Start a new project.
- Add data and create a bookmark.
- Create a 3D scene.

ArcGIS Pro, the latest evolution of the Esri line of GIS software products, is a solution for today's GIS professional. It offers 2D and 3D visualization and analysis within an intuitive, easily navigable interface. ArcGIS Pro seamlessly integrates with networks and the cloud to allow researching, developing, sharing, publishing, and collaborating on GIS projects. The desktop application is designed to be used with ArcGIS Online; for instance, maps authored in ArcGIS Pro can be published to ArcGIS Online, shared with and modified by other users, and then brought back into ArcGIS Pro. You already got an introduction to ArcGIS Online in the first chapter, so now it is time to meet the main character of this book.

ARCGIS PRO SYSTEM REQUIREMENTS

To take full advantage of the 3D capabilities of ArcGIS Pro, you need an adequate system. Complete system requirements can be found in ArcGIS Pro Online Help, under Get Started > Set Up > System Requirements (links.esri.com/SysReqs).

The following short exercises provide an initial overview of ArcGIS Pro. They are designed for brand-new GIS users as well as users who are familiar with other Esri mapping products. You will be introduced to the interface, start exploring some maps, and accomplish some common GIS tasks. These tasks include looking at feature attributes, turning on labels, and modifying map contents. Then, you will work with 3D maps. This chapter will prepare you to successfully complete the GIS project scenarios presented in subsequent chapters.

DATA

The exercise data for this book can be downloaded from ArcGIS Online. Information on the location and installation can be found in the preface section, under "Hardware and software requirements."

In C:\GTKAGPro\World\World.gdb:

- Cities—point features that represent major cities of the world with a population greater than 1 million *(Source: ArcWorld)*
- Countries—polygon features that represent world countries, including demographic data and air pollution estimates *(Source: ArcWorld; US Census International Division, CIA Factbook; The World Bank)*
- Latlong—line features that represent latitude and longitude *(Source: Esri)*
- Ocean—grid of polygons used to display a single-color background behind land features *(Source: Esri)*

In C:\GTKAGPro\3D:

- buildings.shp—polygon building footprints for a section of a Manhattan neighborhood in New York City *(Source: New York City Department of City Planning)*

Exercise 2a: Learn some basics

Estimated time to complete: 40 minutes

In this exercise, you will create a new ArcGIS Pro project, add a map to it, explore map features, and modify the map view.

Exercise workflow

- Create an ArcGIS Pro project using an imported map and a new folder connection.
- Modify map layers by changing visibility, rearranging the order in which layers are drawn.
- Navigate around the map and explore map features and attribute tables.
- Use the Select tool.

Start a new project

1. Start ArcGIS Pro.

At start-up, you are asked to sign in using your ArcGIS Online organizational account. Your login authorizes your ArcGIS Pro license and makes it easier to obtain or share data with ArcGIS Online organizations or an ArcGIS portal. You can choose the Sign Me In Automatically option if you do not want to sign in each time you start the application.

In Settings > Licensing, you can choose to work offline. This option is available if you are working without an internet connection. However, keep in mind that you will not see basemaps or any other data layers that are served on the web.

2. Enter your username and password to sign in.

3. On the opening screen, choose to open a Map template. Give it a unique name (for example, **FirstLook**), and save it to either the default or another location, such as your C:\GTKAGPro folder. Maintain the default option to create a folder, and click OK.

ArcGIS interface elements change periodically to keep up with evolving software functionality, so the images in this book may not exactly match the software interface you see.

PROJECT TEMPLATES

Using a project template (.aptx) is usually the quickest way to start a project. Templates are shareable project packages, including the specific basemaps, connections, datasets, toolboxes, or add-ins that are most helpful to your project. ArcGIS Pro offers some basic templates (catalog, global scene, local scene, and map), and other industry templates are available as well. You can also create your own template: package an existing project and save it to the Documents\ArcGIS\ProjectTemplates folder in your user profile. You can also share it to your organization's network or through ArcGIS Online. Using an organizational template is a good way to ensure consistency between projects.

An ArcGIS Pro project is much more than a map document. A project file (.aprx) or project package (.ppkx—essentially a zipped project file, ideal for sharing) may contain multiple maps, geodatabases, folder connections, layer files, task lists, models, toolboxes, and more. It is an easily shareable container that holds everything you need for your GIS project.

You see a map with a ribbon at the top. The ribbon contains tabs that have various buttons, tool groups, tools, and drop-down options. Notice the elements highlighted in the accompanying graphic of the ArcGIS Pro user interface. All windows can be resized, docked, or detached, or turned off when they are not needed. Tabs and tools change depending on the content you are working with.

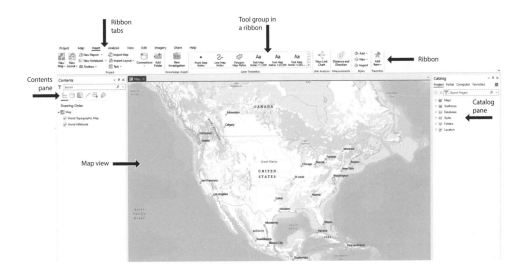

To dock or move a pane (such as Contents or Catalog), click the top of the pane, and press and hold the mouse button. As you begin to move the mouse pointer, the pane becomes transparent, and arrows point to locations where the pane can be docked. Drag the pane to an arrow and then release it to choose the docking location. Or drag and release anywhere if you prefer a detached pane.

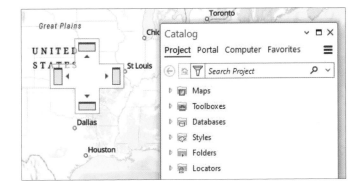

Panes can be stacked to save space. Click the tabs at the bottom of the panes to switch between them.

When a pane is docked, click the Auto Hide button—it looks like a pushpin—to activate tabbed behavior. A tab replaces the pane when you are working in the map view; click the tab to reveal the hidden pane.

4. Arrange or resize the panes as you prefer.

Import a map document

Map documents created with other applications (such as ArcMap™) can be imported into an ArcGIS Pro project.

1. Click the Insert tab, and in the Project group, click the Import Map button.

If your ArcGIS Pro window is small, you may see only the icon.

2. Import World_data.mxd from your C:\GTKAGPro\World folder.

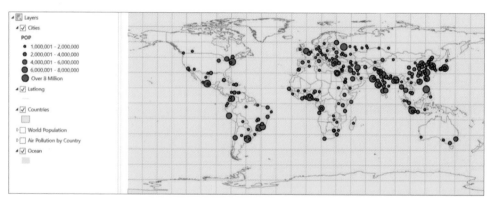

Notice that the Contents pane is populated with the map layers. Some layers are turned on (indicated by a check mark), and others are turned off.

When the map you import has multiple data frames, each frame becomes a separate map in ArcGIS Pro.

Create a folder connection

Creating folder connections for your projects allows you to quickly add data or modify a layer's source data. A folder connection prevents you from having to search through multiple system folders.

1. On the Insert tab, in the Project group, click Add Folder.

2. Browse to C:\GTKAGPro, select the World folder, and click OK.

3. On the Catalog pane, expand Folders to see the new connection.

    There is also a connection to the project folder, which is created by default when you start a new project.

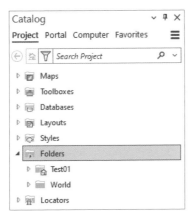

Modify map contents

The Contents pane allows you to modify the map's layers. In the next steps, you will get a quick overview of how to work in the Contents pane.

Look at the Cities layer. Cities are represented by graduated point symbols that correspond to population values—the larger the point, the larger the city's population.

1. Click the check box to the left of Cities to clear it (thereby turning the layer off in the map).

2. In the same way, turn off Countries, and then click the World Population check box to turn it on. Expand the World Population legend by clicking the triangle symbol next to the check box.

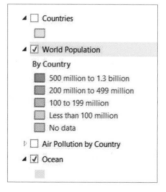

The countries are symbolized by population categories using a graduated color ramp—the darker the color, the higher the population.

If you want to remove a layer from the map entirely and not just turn it off, right-click the layer name and click Remove.

3. Collapse all legends.

Collapsing the legends makes it easier to reorder layers in the Contents pane.

4. Move Latlong below Air Pollution by Country by dragging to move the layer and then releasing it in the desired location.

5. Turn on Air Pollution by Country.

You cannot see this layer because it is underneath World Population. To see the Air Pollution layer, you can turn off World Population or reorder the layers. Think of layers as shapes drawn on sheets of transparent paper. You typically place points and lines above polygons so that they are not covered up.

6. Turn off World Population to examine the Air Pollution by Country layer. Expand its legend so that you can better understand the map.

This layer shows particulate matter (PM) concentration measurements for each country.

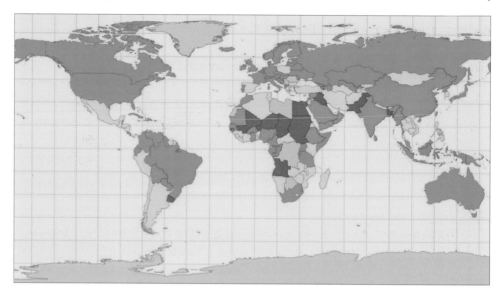

On which continent are PM concentrations highest?

You can find the answers to the questions posed in this book in the book's online resources (links.esri.com/GTKPro3.2).

7. When you finish, turn on World Population again.

You may also collapse the Air Pollution legend again if you want. Next, you will change the map's name from Layers to World.

8. In the Contents pane, click Layers once to highlight it, and click it again to make it editable. Type **World** and press Enter.

Explore the map

To navigate around the map, you will use the Explore tool.

1. On the Map tab, in the Navigate group, make sure the Explore tool is active.

    If you want, you can right-click the Explore tool and click Add to Quick Access Toolbar. A small Explore tool button is added at the top of the application window, so you can easily switch to Explore mode without changing tabs. To undo this change, right-click the button and click Remove from Quick Access Toolbar. You can add any tool you want to the Quick Access Toolbar.

2. Navigate around the map:
    - To zoom in and out, move the mouse while pressing and holding the right mouse button. (You can also zoom using the mouse wheel button.)
    - To pan, press and hold the left mouse button, and drag the map.

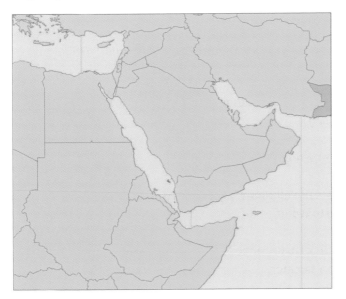

If you are using a touch screen monitor, try panning and zooming using conventional touch screen motions. (Swipe to pan, pinch to zoom out, and reverse pinch to zoom in.)

3. With the Explore tool still active, click any country to see its attributes in a pop-up window. (Zoom and pan as needed.)

Notice that the feature first flashes blue and then flashes red.

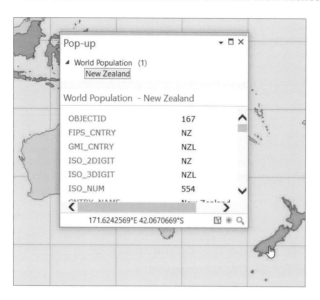

As explained in the first chapter, *attributes* are pieces of information about a geographic feature in a GIS. They are usually stored in a table and linked to the feature by a unique identifier. For example, city attributes might include the city's name, the county in which it is located, total population, and demographic data. Attribute tables also store information about feature geometry, such as length and area (for lines and polygons, respectively).

4. Close the pop-up window.

5. Go to the full extent of the map: on the Map tab, in the Navigate group, click the Full Extent button.

Examine the contextual ribbon

Before you open an attribute table, you should understand that the ribbon and tools you see in ArcGIS Pro are automatically updated according to context. When you are working with the map view, you have the standard available tabs and tools, but when you work with layers, additional contextual tabs and tools appear.

1. Make sure the World map frame is selected in the Contents pane.

    The map frame ribbon contains multiple tabs.

2. Click each tab to see the tools and functions offered for each tab. (If you click Project, click the back arrow to return to the map.)

    If you close the Contents pane, how do you restore it?
    How do you find a geoprocessing tool?

3. In the Contents pane, activate and turn on Cities.

    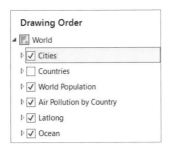

    Notice that the contextual ribbon appears, on which there are three additional tabs that relate to feature layers: Feature Layer, Labeling, and Data.

Examine feature attributes

1. Make sure Cities is still active (highlighted) on the Contents pane. On the contextual ribbon, click the Data tab. Notice the functions that are available here.

2. In the Table group, click the Attribute Table button.

The attribute table opens. It can be docked or floating. The table contains all the attributes of all the features in a layer. Compare this table to the Explore pop-up window, which shows only the attributes of a single feature.

You can also open the attribute table by right-clicking the feature layer in the Contents pane and then clicking Attribute Table.

3. Arrange the table as you prefer, and examine the attribute data.

4. Right-click the POP heading (scroll to the right as necessary), and click Sort Descending.

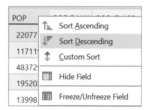

Which city has the largest population?

Select features

1. Click the gray square to the left of the top row.

Notice that the row is highlighted in aqua, and so is the associated feature on the map in Shanghai, China. This step is called selecting a feature. You can *select* one feature at a time or multiple features.

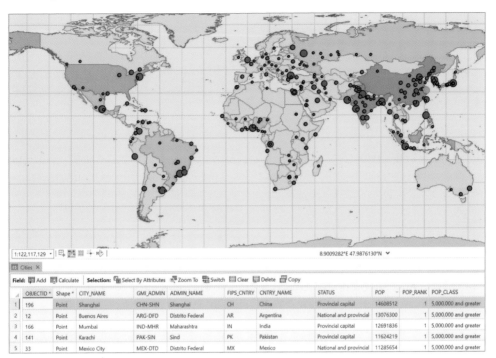

2. Press and hold the Ctrl key and select the five most populous cities in the table.

   Notice that the cities are highlighted on the map, and at the bottom of the table, you can see how many features are selected. By clicking the buttons at the bottom of the attribute table shown in the accompanying graphic, you can choose to show all records in the table or only the selected records.

3. On the Map tab, in the Selection group, click the Clear button.

   This step clears the selection.

4. Close the table by clicking the X (in the upper-right corner if the table is floating, or next to the table name).

You can also select features manually in the map, without using the attribute table.

5. Zoom to any country. On the Map tab, click the Select tool, and click a country. Notice that a button appears. Click the down arrow to see a list of overlapping features.

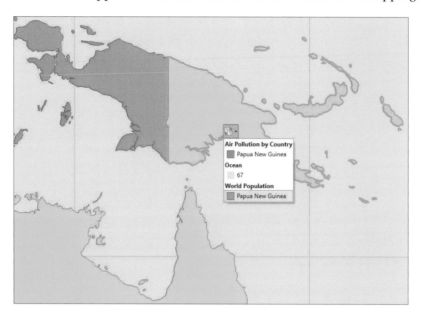

If you open the World Population attribute table, you see that the selected country's row is highlighted in the table. Clicking different layers in the overlapping features list changes the selection and is reflected in the respective feature's attribute table.

To select multiple features, press Shift while clicking each desired feature. Or draw a rectangle (or another shape—click the down arrow on the Select tool in the toolbar to reconfigure the tool's default behavior) to select all features within the shape's extent.

Selections are more than a visual aid; you can generate calculations or run geoprocessing operations on a selected set of features rather than the entire set. You will do this later in the book.

6. To clear the selection, click the Clear button in the Selection group. Click the Explore button again to turn off the Select tool.

7. Zoom to the full extent of the map. At the top of the application, click the Save Project button.

This exercise was meant to get you comfortable navigating the ArcGIS Pro interface. It also underscores the relationship between geographic features and their attributes.

You will continue working with the same project in the next exercise. You can continue to the next exercise or exit ArcGIS Pro and come back to it later.

Exercise 2b: Go beyond the basics

Estimated time to complete: 20 minutes

In this exercise, you will change feature symbols, configure and display feature labels, use the Measure tool, add a cloud-hosted basemap, and package your project to share online.

Exercise workflow

- Customize the appearance of the map using symbols and labels.
- Use the Measure tool to find the approximate distance between cities.
- Examine and add a basemap.
- Create a map package for sharing.

Modify feature symbols

1. Open the ArcGIS Pro project you created in exercise 2a.

2. In the Contents pane, highlight Cities. On the contextual ribbon, click the Feature Layer tab and then click the Symbology button.

   The Symbology pane opens.

   Symbology refers to the way GIS features are displayed on a map. The symbology dictates not only the size, shape, and color of map features but also the conventions and rules used for displaying these features. Symbology is not just for looks; it also conveys meaning to map readers. You have already noticed, for instance, that the Cities layer is represented using *graduated symbols*. In this case, cities are represented by a range of symbols based on an attribute field named POP (which stands for *population*). So, the greater the population, the larger the symbol.

Here are some other ways to access the Symbology pane:

- *Right-click a layer in the Contents pane and click Symbology.*
- *In the layer's legend, click a symbol icon.*
- *If the Symbology tab is already visible (stacked with the Catalog pane, for instance), simply highlight the layer in the Contents pane, and then click the tab.*

GIS IN THE WORLD: SKI PATROL USES ARCGIS TRACKING ANALYST FOR RESCUE MISSIONS

Colorado's Winter Park Ski Resort has successfully used GIS and GPS technology to track lost skiers and snowboarders. Lost resort visitors can be tracked using their GPS-enabled smartphones. GPS points are imported into ArcGIS software, and then the rescue team's locations are tracked using the ArcGIS Tracking Analyst extension so that the mission can be monitored in real time. Read more in *ArcWatch*, "GIS to the Rescue": www.esri.com/esri-news/arcwatch/0114/gis-to-the-rescue.

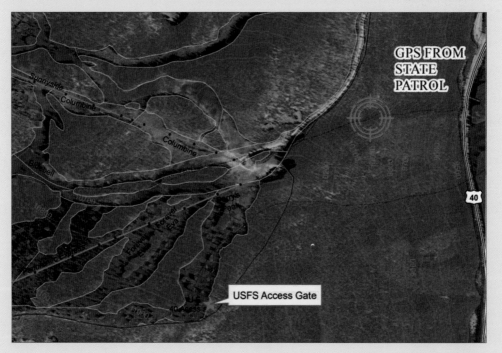

FEATURE SYMBOL OPTIONS

Spatial data, in its simplest form, is a collection of points, lines, and polygons. But you can modify default symbols to make your map more readable and informative. You can display map features in the following ways:

- Single symbol—one symbol is used for all features in a layer.
- Unique values—used for categorical data, different symbols represent various attributes. For example, imagine you have a polygon layer of parklands, and it has an attribute that identifies each feature as either a state or national park. You can symbolize state parks using a black outline and green fill and national parks using a gray outline and brown fill, for example.
- Graduated colors—used for quantitative data, different colors represent different value ranges. Often, a color ramp is chosen in which darker colors correspond to higher values. For example, in this map, the World Population layer is symbolized using graduated colors, where darker colors represent higher population attribute values. The user decides how values are classified and how many categories there should be.

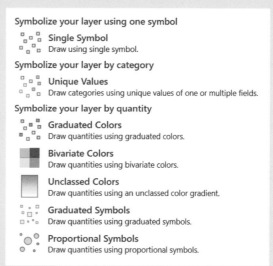

- Graduated symbols—used for quantitative data, symbols increase in size with increased values. Earthquake epicenters might be given graduated symbols that represent magnitude values. The user chooses the classification method and number of categories for this option.

3. On the Symbology pane, click the symbol template to edit it.

4. In the Symbology pane, on the Gallery tab, expand ArcGIS 2D, and choose any 2D symbol template you like from the Gallery, such as Circle 5.

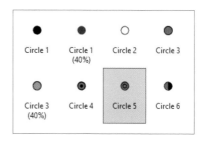

5. Close the Symbology pane (click the X in the upper-right corner).

Label features

You will add labels to your map to make it more informative.

1. In the Contents pane, right-click Cities, and click Label.

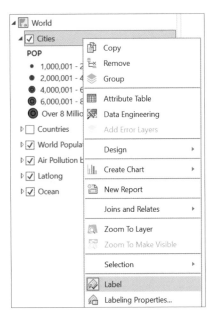

City labels appear cluttered at the full extent of the map. You will set a visibility scale so that city labels appear only when zoomed in.

2. With Cities still highlighted in the Contents pane, click the Labeling tab on the contextual ribbon.

Notice that the labeling field is set to the CITY_NAME attribute. You can change this field so that the labels display city population or some other attribute if you want.

3. In the Visibility Range group, next to Out Beyond, type **1:75,000,000** and then press Enter on the keyboard.

This range means that when you are zoomed to full extent, you do not see the labels, but as you zoom in to a scale of 1:75,000,000 or closer, labels are turned on.

You can also choose a predefined extent from the drop-down menu.

4. Zoom in until the labels appear.

ON YOUR OWN

Turn on labels for the World Population layer. Modify the default visibility scale if you want.

Measure distances

What if you want to know the approximate distance between Lima, Peru, and Rio de Janeiro, Brazil?

1. Make sure that the Explore tool on the ribbon is activated.

2. Zoom to South America. (Ensure that you can see Lima and Rio de Janeiro.) On the Map tab, in the Inquiry group, click the Measure tool down arrow. Notice the different configurations that you can choose for this tool. Click Measure Distance.

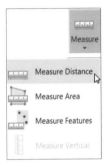

3. In the Measure Distance window, choose Miles from the list on the right, which provides options for units.

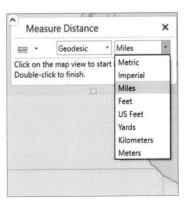

4. Click the point that represents the city of Lima, and then double-click the point that represents Rio de Janeiro.

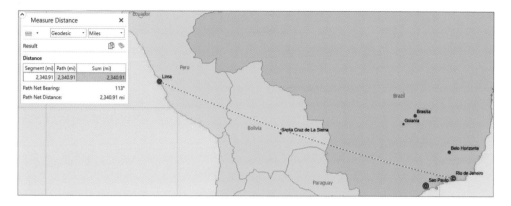

Notice that the tool provides the distance measurement for the current segment as well as the entire path length if you click multiple points.

5. Click the Clear Results button in the Measure Distance window to clear the previous measurement and start again.

ON YOUR OWN

Measure distances between other cities on the map.

The Measure window remains open as long as the Measure tool is active. Choosing any other tool closes it.

6. Click the Explore tool.

Add a basemap

Esri provides several basemaps that you can use in any project. Basemaps are like dynamic graphics of paper maps or imagery. They do not have attributes, but they provide a professional appearance to your map and visual context for your study area. They cover the entire world, so no matter where or at what scale your project is, you can use a basemap. Depending on your project, you can choose to use satellite imagery or a topographic or street map. The basemaps provided by Esri were created collaboratively by several organizations. An organization can create a custom basemap that is used for all its projects. Or you may decide to not use a basemap at all.

1. Zoom to the full extent of the map.

> **REMIND ME HOW**
>
> Click the Full Extent button.

2. Turn off the Latlong and Ocean layers.

3. On the Map tab, in the Layer group, click the Basemap button. From the basemap gallery, select Oceans.

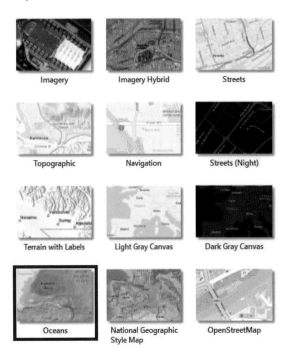

Depending on the size of your window, you may need to zoom in slightly to see the new basemap. Basemaps are hosted by ArcGIS Online. You will not be able to load one if you do not have a good internet connection.

When you zoom in, you may notice duplicate city labels. To avoid double labeling, you can turn off either your Cities labels or World Oceans Reference, a reference layer that comes with the Oceans basemap. Turning off this layer does not affect the basemap, only the labels and political boundaries. To see this reference layer, click the List by Data Source button at the top of the Contents pane.

4. To remove the old Ocean background layer, in the Contents pane, right-click Ocean, and click Remove.

Package and share the map

1. Zoom to the full extent, and then save the project by clicking the Save button in the upper-left corner of the window.

2. On the Share tab, in the Package group, click the Map button.

If you have other project elements you want to share—such as styles, toolboxes, task lists, or attachments—sharing a project package, which includes these other elements, is more appropriate. (You will learn more about styles, toolboxes, and task lists in subsequent chapters.) In this case, sharing a map package is the best choice.

3. In the Package Map pane, maintain the default parameters and then click Analyze. When Analysis is complete, click Package.

To package a map, there must be a description in the map's metadata. To access the metadata, in the Contents pane, right-click the map title, click Properties, and go to the Metadata tab. Another way to edit the metadata is to right-click the map title and click Edit Metadata. There must also be a summary and tags (keywords for searching), which are read from the metadata; if the summary and tags are not in the metadata, you may enter them in the Package Map pane.

ON YOUR OWN

Go to www.arcgis.com, and sign in to your account. Click Content and then find the map package under My Content. Click the map and share it with any group or individual you want.

If you are not continuing to the next exercise, you may exit ArcGIS Pro.

Exercise 2c: Experience 3D GIS

Estimated time to complete: 20 minutes

Presenting a 3D map engages your audience and provides a wow factor that your 2D maps may lack. But it is not just about looks—with the power of 3D GIS, you can visualize and modify skylines, experience realistic topography, and perform 3D analysis. For example, you can construct sightlines in hilly terrain. In this exercise, you will learn how to convert a 2D map to 3D.

Exercise workflow

- Start a new project, and add a layer of polygons that represents buildings in New York City.
- Use the Bookmark tool.
- Convert a 2D map to a 3D map, and then extrude features based on the building height attribute to visualize buildings in a more realistic perspective.

Start a new project

First, you will start a local scene project.

1. In ArcGIS Pro, if you are starting from an existing project, click the New Project button at the top left of the window. If you are starting a new ArcGIS Pro session, choose the Local Scene template.

2. Name the project **3D**, and store it in the C:\GTKAGPro\3D folder. Clear the option to create a new folder for this project because a folder is provided for you.

3. In the 3D project, on the Insert tab, click the New Map down arrow, and click New Map.

ArcGIS Pro modifies the New Map default button to reflect your previous choice, so sometimes you will click only the button and not need the arrow.

Add data and create a bookmark

Next, you will add a shapefile of some buildings in New York City.

1. In the Catalog pane, expand Folders and then expand 3D.

The project automatically stores a folder connection in the location in which you saved your project.

To create additional folder connections, right-click Folders and click Add Folder Connection.

Notice that the 3D folder contains a geodatabase and a toolbox, both of which are currently empty and were created automatically when you created the project. (You will learn more about what these things are in subsequent chapters.) The folder also contains a GIS data file named buildings.shp, which was already there.

2. In the Catalog pane, click to highlight buildings.shp, and then drag and release it in the map view.

Building footprints that span 10 blocks of New York City's Upper West Side neighborhood are added to the map, and the map display zooms to the extent of the new layer. Colors will vary because colors are applied randomly when data is added to a map.

When data resides in the project folder, it may be the easiest place from which to drag data to the map, as you did here. If the data resides elsewhere, either create a new folder connection or use the Add Data button, browse to the desired dataset, and click Select. Notice that you can configure the Add Data button to add different types of data (for example, route events or address layers) by clicking the down arrow.

Next, you will create a bookmark of the buildings layer extent.

3. On the Map tab, in the Navigate group, click the Bookmarks button, and click New Bookmark. Name it **UWS Buildings**, and then type a brief description: **Set of buildings in New York City's Upper West Side neighborhood**. Click OK.

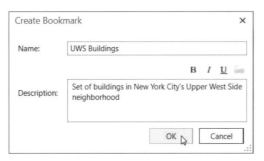

Now when you click Bookmarks, you see a thumbnail along with the name of the bookmark.

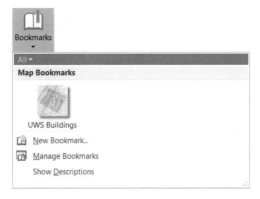

Create a 3D scene

Now you will convert a 2D map into a 3D scene (without losing your 2D map) so that you can visualize the heights of the buildings in the map. ArcGIS Pro basemaps can be used for both 2D and 3D maps. You can turn your buildings layer into a 3D layer because it includes a roof height attribute, measured in feet.

1. In the Contents pane, right-click the buildings layer, and click Attribute Table. Scroll to the last column, right-click the Height attribute, and sort in descending order.

What is the height of the tallest building in this layer?

---

2. Close the attribute table.

3. On the View tab, click the Convert down arrow and click To Local Scene.

When the conversion is complete, a new 3D scene is added to the project. Now you will enable 3D rendering for buildings.

4. In the Contents pane, drag the buildings layer to 3D Layers.

5. In the Contents pane, right-click the buildings layer, and click Zoom To Layer.

Do not worry if the buildings do not render correctly in the 3D map—they must be extruded before they can be displayed properly. *Extrusion* is the process of stretching flat 2D features vertically so that they appear three-dimensional. You will extrude the buildings based on the building height attribute field.

6. Make sure that the buildings layer is selected in the Contents pane. Click the Feature Layer contextual tab. In the Extrusion group, click the Type down arrow, and click Base Height. For Field, click HEIGHT, and for Unit, click Feet.

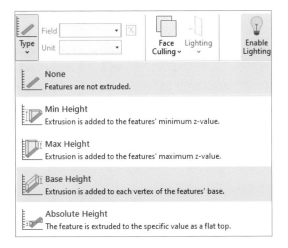

You are not limited to extruding height measurements. You can create a 3D population map by extruding a population value. Any numeric value can be displayed in 3D.

To learn about the different extrusion types, go to the ArcGIS Pro Help topic: Maps and scenes > Layers > Layer properties > Extrude features to 3D symbology.

7. On the Feature Layer contextual tab, click the Symbology tool. In the Symbology pane, click the current symbol, expand the ArcGIS 3D style set, and click Concrete.

8. Click the Explore tool (on the Map tab of the ribbon, or on the Quick Access Toolbar if you placed it there). Point to the Explore tool and familiarize yourself with navigation methods.

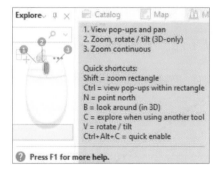

You can click the scroll wheel button on the mouse to tilt and rotate a 3D scene.

9. Navigate the 3D scene. Use the mouse wheel button to experiment with zooming, tilting, and rotating the scene.

What if you want to present a 2D map alongside a 3D map? You will do that next.

10. At the top of the map, click the Map_3D tab and drag it to place it alongside or below the 2D map.

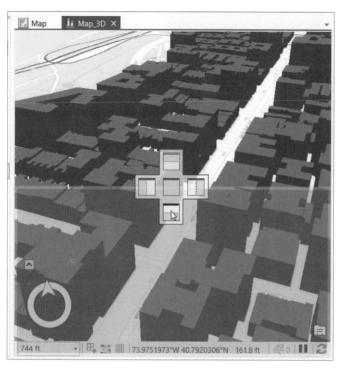

11. On the View tab, in the Link group, click the Link Views down arrow, and then click Center and Scale.

Now when you navigate in one map, your view will adjust accordingly in the other map because the views are linked.

12. Save and close the project.

If you are not continuing to chapter 3, you may exit ArcGIS Pro.

Summary

These exercises gave you a first look at ArcGIS Pro. You learned how to import existing ArcMap map documents, perform some common GIS tasks, add a basemap and local data, and share a map. You also practiced converting 2D data to 3D.

At this point, you should begin to feel comfortable working with ArcGIS Pro, because it is time to move on. Coming up: integrating desktop and cloud GIS, data investigation, using symbology to reveal data patterns, creating data from scratch, analyzing time and surface variations, deriving new information from existing data, and learning map presentation tips. As you work through real-world scenarios touching many industries (for example, health care, conservation, city planning and maintenance, vineyard management, and more), you will develop a better understanding of what ArcGIS Pro can do. You will learn how to use ArcGIS Pro to streamline operations, enhance workflows, reveal new information, and solve problems in your own community or workplace.

Glossary terms

select
symbology
graduated symbols
extrusion

## CHAPTER 3

Exploring geospatial relationships

Exercise objectives

3a: Extract part of a dataset
- Add data to a project.
- Select features by attributes.
- Export the selection to a new dataset.

3b: Incorporate tabular data
- Join data tables.
- Apply informative symbols.
- Import layer symbology.
- Use the Swipe function to compare layers.
- Overlay additional data.

3c: Calculate data statistics
- Add a new field.
- Calculate field values.
- Display a new field.
- Calculate summary statistics.
- Examine infographics.

3d: Connect spatial datasets
- Relate tables.
- Spatially join data.

The power of GIS extends far beyond exploring digital maps. You can combine datasets, enrich them withĬ new attributes, derive statistics from them, and obtain new information based on their relationships. In this chapter, you will begin taking advantage of some of the more sophisticated capabilities offered by ArcGIS Pro.

Scenario: You have been hired by a state health coalition that is focusing its efforts on raising awareness about and lowering obesity rates in the state of Illinois. Obesity prevalence has risen dramatically throughout the state over the past decade, along with the incidence of type 2 diabetes, but the prevalence of obesity is higher in some areas than others. You will explore obesity prevalence rates by county, create visual aids for displaying a year-over-year rising trend, and begin to explore possible reasons why some counties may have higher prevalence rates than others. For instance, is there a link between average income level and obesity levels? What about comparing obesity rates to data that shows limited access to grocery stores—are these "food deserts," with lesser accessibility to healthy foods, more likely to have higher obesity rates?

Your job is to combine obesity statistics provided by the Centers for Disease Control and Prevention (CDC) with county GIS data to do an initial visual analysis of the data, which will eventually lead to more rigorous analysis techniques. The end objective of the analysis is to create informative products that will educate and guide advocates and policymakers.

GIS IN THE WORLD: AN INTERACTIVE ATLAS OF HEART DISEASE AND STROKE INCIDENCE

The Interactive Atlas of Heart Disease and Stroke (https://nccd.cdc.gov/DHDSPAtlas), created by the CDC and powered by GIS, creates a visual guide for anyone who wants to see where cardiovascular disease occurs and offers insight about who is at highest risk. Read more about the project in *ArcNews*, "CDC Aims at Reducing Annual Heart Disease- and Stroke-Related Deaths in the US": www.esri.com/about/newsroom/arcnews/cdc-aims-at-reducing-annual-heart-disease-and-stroke-related-deaths-in-the-us.

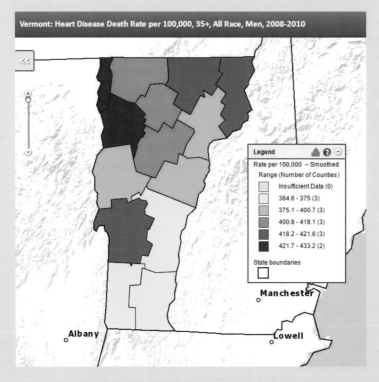

Figure 3.1. The CDC atlas allows you to apply various filters to get the information you need. Here, the map shows heart disease death rates in Vermont between 2008 and 2010, limited to men above the age of 35. This map was created using the Interactive Atlas of Heart Disease and Stroke, a website developed by the Centers for Disease Control and Prevention, Division for Heart Disease and Stroke Prevention. http://nccd.cdc.gov/DHDSPAtlas.

DATA

In the C:\GTKAGPro\HealthStudy\Data folder:

- IL_food_deserts.shp–a shapefile containing polygons that represent low-income areas with limited access to grocery stores *(Sources: Data and Maps for ArcGIS; USDA)*
- IL_med_income.shp–a shapefile representing Illinois counties with an attribute for median income values *(Sources: Data and Maps for ArcGIS; US Census Bureau)*
- Obesity_Prevalence.dbf–a spreadsheet containing obesity prevalence rates for Illinois counties from 2004 to 2010 *(Source: Centers for Disease Control and Prevention)*
- us_cnty_enc.shp–a shapefile containing polygons that represent the county boundaries for the East North Central region of the midwestern United States *(Source: Data and Maps for ArcGIS)*

Exercise 3a: Extract part of a dataset

Estimated time to complete: 20 minutes

In this exercise, you set up your project. You will add data to a new project and then do some basic processing to get your data in shape. Once you have completed these steps, you will be ready for a deeper exploration of your data.

Exercise workflow

- Open a project and add a shapefile of polygons that represent counties from five states in the East North Central region of the Midwest.
- Select the features with a STATE_NAME attribute of Illinois.
- Export the selected features to a new dataset that contains counties in Illinois only.

Add data to a project

1. Start ArcGIS Pro, click Open Another Project, find the C:\GTKAGPro\HealthStudy folder, and open HealthStudy.aprx.

    The project opens with a 2D map that includes the Esri topographic basemap centered on North America.

ArcGIS interface elements change periodically to keep up with evolving software functionality, so the images in this book may not exactly match the software interface you see.

2. On the Map tab, ensure that the Explore tool is active, and click any area in North America.

   Notice that there is no pop-up window. That is because basemaps function the same way as images—they do not have individual features with attributes.

3. In the Catalog pane, expand the HealthStudy\Data folder.

   Notice that this folder contains three shapefiles (.shp), two layer files (.lyrx), and one database file (.dbf).

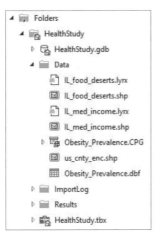

4. Press and hold Ctrl, and click us_cnty_enc.shp and Obesity_ Prevalence.dbf, and then drag them to the map display.

   Your map project now includes a shapefile that contains county features for the East North Central region of the midwestern United States. The area includes five states: Wisconsin, Michigan, Illinois, Indiana, and Ohio. The project also includes a nonspatial .dbf table, listed in the Contents pane under Standalone Tables. Because the table is tabular and does not have spatial attributes, it cannot be displayed on the map in this form.

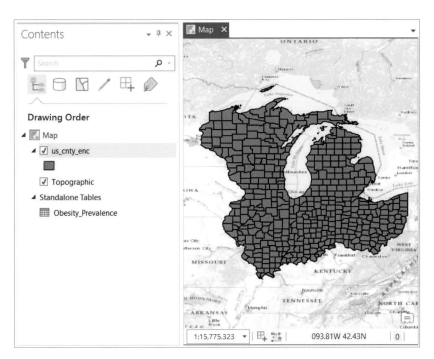

The colors will vary because they are randomly assigned.

5. With the us_cnty_enc.shp layer selected in the Contents pane, click the Data tab, and then click the Attribute Table button. Examine the table.

What is the field name that indicates the state within which the county features are located? How many residents of Wayne County are between the ages of 22 and 29 years?

---

6. Close the attribute table.

You'll want to limit the us_cnty_enc.shp dataset to the state of Illinois only. You can consider a few operations to achieve this subset:

Definition query. You can set a definition query to limit the visible counties to only those within the state of Illinois. Definition queries are helpful when you want to work with a subset of data in a map while maintaining the source data. You can set a definition query on the Data tab of the contextual ribbon. However, you must think ahead—what is next in the workflow? You want to append Illinois obesity statistics (which are in a stand-alone table) to a spatial layer of Illinois counties. You will do this by performing a join operation, which is explained in detail in exercise 3b. But with a definition query defined, the source data does not change, so the operation will attempt to join Illinois obesity rates to every county in the five-state dataset. Although this operation works, it takes a little longer, so it is best to go another route.

You will work with definition queries later in the book.

Clip. You can use the Clip operation to select a portion of us_cnty_enc.shp based on another layer—namely, a layer that provides a boundary of the state of Illinois. However, you do not have an Illinois boundary layer now. There is another way.

You will work with the Clip tool later in the book.

Select and export. You can select only those counties that belong to the state of Illinois and export the selected features to a new dataset. Then you will have an exact one-to-one correlation between your spatial attribute table (because the new output will have one record for each Illinois county) and your stand-alone table (obesity statistics). This method is the best for the current study.

Select features by attributes

You will create a selection in which the state name attribute value equals Illinois.

1. On the Map tab, open the Select By Attributes tool.
   The Select By Attributes tool opens.

2. For Input Rows, click the us_cnty_enc shapefile. For Selection Type, maintain the default New selection (click the down arrow to see additional options).

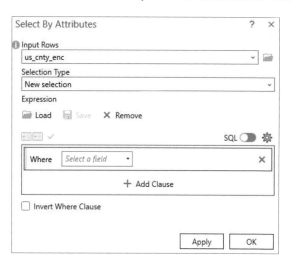

3. Click Add Clause, and then choose the correct options to create the following expression: **Where STATE_NAME is Equal to Illinois**.

The expression you created is called an *attribute query*—a request for features in a table that meet user-defined criteria.

4. Click the red X to delete any unused fields in the definition query, and click OK. Illinois counties are now highlighted in the selection color (cyan, by default).

Now you will create a feature layer from the selected set of counties.

You have made a selection based on attributes, but you can also make a selection based on location. For instance, suppose you want to select parcels that are within the boundaries of a flood zone. In this case, use the Select By Location tool. You can choose from many relationship types, including "Completely within," "Have their center in," "Intersect," and "Within a distance." You will use Select By Location later in the book.

Export the selection to a new dataset

1. In the Contents pane, select us_cnty_enc.

2. On the Data tab, in the Export group, click the Export Features button.

The Export Features tool pane opens, in which you can modify parameters and environments.

3. In the Export Features pane, for Input Features, click us_cnty_enc. For Output Feature Class, maintain the default location but change the Output Name to **Illinois_cnty**.

To see the location of the output feature class, click inside the Output Feature Class box and scroll to the far right. After you have typed the output name, click outside the Output Feature Class box to confirm the output name.

Note that the new output will be a geodatabase feature class rather than a shapefile. By default, all processing outputs are saved to a project geodatabase—the conversion from shapefile to geodatabase feature class is done behind the scenes.

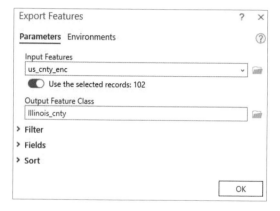

4. Click OK.

5. Remove the old shapefile us_cnty_enc from the map contents. Zoom to the extent of the Illinois_cnty layer. Change the map name to **Illinois**.

REMIND ME HOW
- Right-click layer and click Remove.
- Right-click layer and click Zoom To Layer.
- In the Contents pane, click Map once to select it and click again to make it editable. Type **Illinois**, and then press Enter.

6. In the Contents pane, click the polygon symbol below Illinois_cnty to open the Symbology pane.

7. In the Symbology pane, click the Properties tab to go to the current symbol properties so that you can modify the symbol manually.

8. Click the Color selection down arrow, choose a light blue, and click Apply.

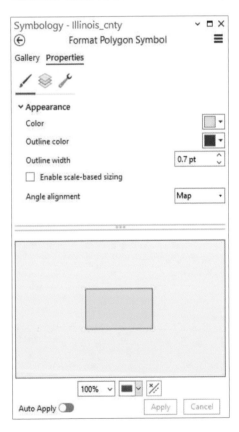

Your preprocessing work is complete—you have created a project and added two datasets: a nonspatial table with obesity statistics for the state of Illinois and a spatial dataset that includes county demographic statistics for five states. Because you are interested in only the state of Illinois, you selected counties with a STATE_NAME field value of Illinois, and you exported these features to a new feature class (Illinois_cnty) stored in the project's geodatabase. In the next exercise, you will join the obesity statistics to the Illinois_cnty attribute table, so you can see the obesity data on a map.

9. Close the Symbology pane and save the project. If you are asked whether you want to save your project in a current version of ArcGIS Pro, click Yes.
    You will continue working with the same project in exercise 3b.

Exercise 3b: Incorporate tabular data

Estimated time to complete: 40 minutes

As you began researching obesity rates in Illinois, you quickly found a CDC website that offers reliable statistics on obesity prevalence. However, the data is offered in a spreadsheet table, and it cannot be displayed on a map because there is no coordinate information. The data is broken down by counties, which is helpful. Using an *attribute join* operation, you can append the spreadsheet table (the *join table*) to your existing attribute table (the *input table*), provided you have a common attribute field in each table—for instance, a feature name or numerical identifier. Then you can visualize, symbolize, label, and analyze the layer based on the new attributes. This is what you will do here: using a county identification code as the common attribute, or *key*, you will join the CDC data to the Illinois county feature class.

> *The attribute field names do not have to be identical—for example, you can join a table with the attribute field name County to another table with an attribute field name Cnty. It must be the same data type—for example, text fields are joined to text fields.*

Exercise workflow

- Join CDC nonspatial data to the spatial layer of Illinois counties so that you can display the CDC data on the map.
- Symbolize the Illinois county layer to reflect the obesity prevalence statistics for a single year (2004).
- Copy the county layer, rename it, and import the symbology scheme used for the 2004 layer, but instead display the 2005 obesity prevalence statistics. Do the same for every year up to and including 2010.
- Use the Swipe tool for a quick visual comparison.
- Add a median income layer and look for data correlations.

Join data tables

1. In ArcGIS Pro, continue working with HealthStudy.aprx in the C:\GTKAGPro\
HealthStudy\Data folder.

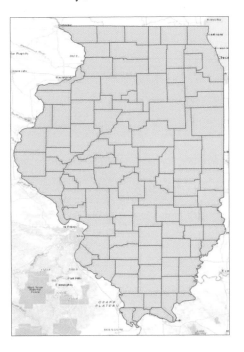

*If you did not complete or save your work from exercise 3a, open HealthStudy.
aprx and add Illinois_cnty from HealthStudy\Results\HealthStudy2.gdb and
Obesity_Prevalence.dbf from the HealthStudy\Data folder.*

2. In the Contents pane, right-click Obesity_Prevalence and click Open.

	OID	State	FIPS_CODE	County	Number04	Percent04	Number05	Percent05
1	0	Illinois	17001	Adams County	11970	24.4	12430	25.2
2	1	Illinois	17003	Alexander County	1673	25.1	1729	26.7
3	2	Illinois	17005	Bond County	3302	24.2	3548	26
4	3	Illinois	17007	Boone County	8807	25.9	9280	26
5	4	Illinois	17009	Brown County	1334	24	1396	25.2

Tables are composed of rows and columns. Columns are often called fields. The column
headers tell you what kind of information is provided for each feature represented by a
row—in this case, the features are Illinois counties. Fields include OID (sometimes called
Object ID), a unique identifier assigned to every row in an ArcGIS table; State, which shows

the state name in which each county is located (in this table, all values are Illinois); FIPS_ Code, a Federal Information Processing Standard (FIPS) code assigned to every US county; County, showing county names; and obesity statistics fields—for example, Number04 shows the number of people who are considered obese in the year 2004, Percent04 shows the percentage of the total county population that is considered obese in 2004, and so on.

3. Scroll to the right of the table to view all the fields.

How many years of data are represented in the table?

PREPARING TABULAR DATA

Sometimes you must do some work to get tabular data in shape. When we first downloaded this table from the CDC website (www.cdc.gov), it looked like the following graphic.

data_Illinois [Compatibility Mode]								
	A	B	C	D	E	F	G	H	I
1									
2									
3	**Obesity Prevalence**							**2004**	
4	**State**	**FIPS Code**	**County**	**Number**	**Percent**	**Lower Confidence Limit**	**Upper Confidence Limit**	**Age-adjusted Percent**	**Age-adjusted Lower Confidence Limit**
5	Illinois	17001	Adams County	11970	24.4	20.3	29.2	24.3	20.0
6	Illinois	17003	Alexander County	1673	25.1	20.6	30.3	25.0	20.3
7	Illinois	17005	Bond County	3302	24.2	19.8	29.5	24.2	19.8
8	Illinois	17007	Boone County	8807	25.9	21.2	31.4	25.7	21.1
9	Illinois	17009	Brown County	1334	24.0	19.1	30.0	24.3	19.4
10	Illinois	17011	Bureau County	6008	23.1	19.1	27.7	22.9	18.7
11	Illinois	17013	Calhoun County	941.3	23.8	19.1	29.0	23.6	18.9
12	Illinois	17015	Carroll County	2982	24.3	19.8	29.5	24.1	19.4
13	Illinois	17017	Cass County	2418	23.9	19.3	29.3	23.8	19.2
14	Illinois	17019	Champaign County	30420	22.8	19.1	26.9	23.0	19.5
15	Illinois	17021	Christian County	6340	24.0	19.7	29.1	23.9	19.5
16	Illinois	17023	Clark County	3018	23.9	19.3	29.1	23.7	19.0
17	Illinois	17025	Clay County	2627	24.8	20.2	30.0	24.7	20.0
18	Illinois	17027	Clinton County	6462	24.0	19.6	29.2	24.0	19.6
19	Illinois	17029	Coles County	9013	23.1	18.9	28.0	23.4	19.2
20	Illinois	17031	Cook County	818500	21.5	20.2	22.8	21.4	20.1
21	Illinois	17033	Crawford County	3750	24.6	20.2	29.7	24.5	20.0

This table was originally formatted for easy reading; however, the table does not work if you try to use it in a table join operation in ArcGIS. The first row of a table must contain the field names that describe the values in their respective columns. So, we had to delete the first three rows and modify the headings in the fourth row. To maintain the descriptive table name "Obesity Prevalence" found in row 3, we incorporated it in the table name, replacing the original, ambiguous file name of data_Illinois. To maintain the years for which the data was collected, we added the appropriate year to each field name that contained statistical values. Then we truncated field names and replaced spaces with underscores (_) to maintain the database rules of no spaces and a limit of 10 characters. We also deleted columns that were not necessary for our study. Finally, we converted the table to an ArcGIS Pro-compatible .dbf file.

To join a spatial table and a nonspatial table, you must ensure that they have at least one field in common—the field name need not match, but the values within the fields must correlate. In this case, the common field is the one that stores FIPS codes. Although the field names can be different, in this case, they have the same logical name: FIPS_Code in the stand-alone table and FIPS_CODE in the attribute table for Illinois_cnty.

4. In the stand-alone table, notice the field named FIPS_Code. If you want, open the Illinois_cnty table and notice that it has a FIPS_CODE field. When you are finished, close both tables.

5. In the Contents pane, select Illinois_cnty. On the Data tab, click the Joins button, and then click Add Join.

The Add Join tool pane opens. In the pane, the Input Table should already be populated with Illinois_cnty.

6. For Input Field, click FIPS_CODE. For Join Table, click Obesity_Prevalence. For Join Field, click FIPS_Code.

Notice the warning symbol (the exclamation point in the yellow triangle) next to Input Field. It tells you that performance will be improved if you first index the table. You can ignore the warning in this case.

7. Run the Add Join tool, and then open the attribute table for Illinois_cnty to see the operation's results. Scroll through all the fields in the table.

Notice that when you pause over a field name, the table name is shown in parentheses, so it is easy to tell which attributes are from the original layer and which are from the joined table.

Number06	Percent06	Number07	Percent07	Number08	Percent08
1936	26.2				7.1
6019	26.3				7.7
11810	27				5.9
1659	27.1				7.7
9561	25.3				27
3779	26.5				3.3
117400	23.8	118700	24.2	122600	24.8

Percent06 (Obesity_Prevalence.Percent06)

Type: Double (15, 6)
Default: <Null>
Read-Only: No
Nullable: Yes
Indexed: No
Required: No

8. Close the attribute table.

Joining data is a powerful GIS function. It is important to realize that the joined attributes remain with the layer you joined them to in the map; however, the underlying feature class is not modified. If you add another copy of the Illinois_cnty feature class to the map, you will not see the joined attributes in the new layer. Because of this, join operations are considered temporary. However, if you save your work, the joined data will be there each time you open the project. If you want to create a dataset that permanently includes the joined attributes, you can now export the join layer to a new feature class with a new name, as you did in exercise 3a.

Apply informative symbols

Now you are ready to display the joined information on the map. You will do so using *graduated colors*. The term *graduated colors* means that instead of using the same symbol for each feature, features are assigned a color that represents a quantity.

1. On the Feature Layer contextual tab, click the Symbology button down arrow.

Make sure Illinois_cnty is still selected in the Contents pane to reveal the contextual tabs on the ribbon.

Notice the symbology options (for more information, see the "Feature symbol options" sidebar in chapter 2). You can choose categories if you want to display descriptions, such as a land-use category, or a value indicator such as Low, Medium, or High. In this case, you want to display counties using a quantity—namely, percentage value.

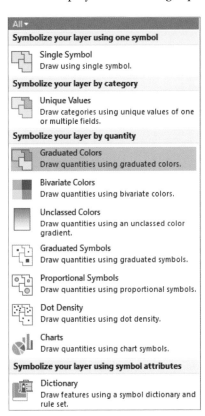

2. In the Symbology drop-down list, click Graduated Colors.

ArcGIS Pro automatically applies a graduated color scheme based on the first numeric field it finds. You will change the field to the one that contains the data that you are interested in.

3. In the Symbology pane, click the Field down arrow and then click Percent04.

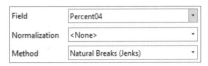

ArcGIS Pro, by default, chooses five classes based on a natural breaks (Jenks) classification method.

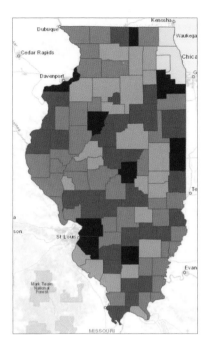

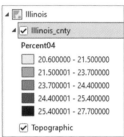

You want to create your own categories to compare maps for every year for which you have data. If you are going to compare values, they must be broken into the same ranges, or your comparison will not be helpful.

4. In the Symbology pane, change the number of classes to **4**. Choose the yellow-to-red color ramp. (If necessary, check the Show Names check box in the Symbology pane to see the names of all available color ramps, scroll through the list, and choose the desired option.)

CLASSIFICATION METHODS

When mapping quantitative data using graduated colors or symbols, you want to classify the data appropriately so that the information is not skewed or misleading. ArcGIS has seven classification methods.

Manual interval classification

Using manual intervals, you can modify the classification breaks manually.

Equal interval classification

With equal intervals, the entire range of the data is equally divided by the number of classes you choose. If the range of values is 1 to 1,000 and there are five classes, this method creates classes of 1-200, 201-400, 401-600, 601-800, and 801-1,000. This method is useful for highlighting changes in the extremes and is most appropriate for familiar data ranges, such as percentages or temperatures. Using this method, it is also possible to have classes that contain no values.

Defined interval classification

A defined interval is similar to an equal interval. However, you can define the interval size to determine the number of classes. If the students in your class have a range of test scores from 50 percent to 90 percent, for example, and the interval size is 10, this method creates four classes: 50-60, 61-70, 71-80, and 81-90.

Quantile classification

With quantiles, all classes have the same number of features. This method is appropriate when data is linearly distributed. It can create a balanced-looking map because no classes have too many, too few, or no values, but the resultant map can be misleading. The key is to increase the number of classes to prevent data from being placed in a different class. When an appropriate number of classes is used, the data can be visually represented more accurately.

Natural breaks (Jenks) classification

Natural breaks classes are based on natural groupings inherent in the data, and boundaries are set in spots in which there are relatively large gaps between values. Developed by Professor George Jenks of the University of Kansas, this method is useful for classifying unevenly distributed data, such as population.

Geometric interval classification

The geometric interval classification scheme creates class breaks based on class intervals that are part of a geometric series, such as a logarithmic distribution. This method was specifically designed to accommodate continuous data and minimize variance within classes to ensure that each class range has approximately the same number of values and that the change between intervals is consistent.

Standard deviation classification

Data that you know is normally distributed (in a bell-shaped curve) can benefit from standard deviation classification. This method creates classes according to a specified number of standard deviations from the mean value. A divergent color scheme is recommended for this type of classification.

5. On the Classes tab in the Symbology pane, click the last row's upper value twice to make it editable, replace the existing value with **35**, and press Enter.

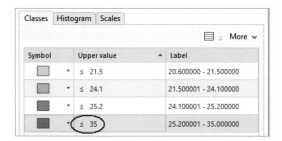

6. Working from the bottom up, enter the following upper values for the remaining three symbol categories: **31**, **27**, **23**. Then edit the symbol category labels to **23% or less**, **23.1% to 27%**, **27.1% to 31%**, and **31.1% to 35%**.

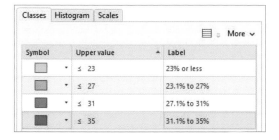

Notice that the new class breaks change your map dramatically: most counties are now in the second-lowest category (23.1% to 27%), because no more than 27 percent of their populations are considered obese. The highest category (31.1% to 35%) is reserved for counties with more than 31 percent of their population considered obese. The higher thresholds are required to accommodate the higher percentages of subsequent years, as you will soon see.

Import layer symbology

Next, you will create additional layers using the same symbology scheme that will reflect data from the years 2005 through 2010.

1. In the Contents pane, right-click Illinois_cnty and click Copy. Then right-click Illinois (the map name) and click Paste.
   Now you have two identical layers in your map.

2. To keep the layers straight, rename the first one **2004**, and the second one (the one at the top of the list) **2005**.

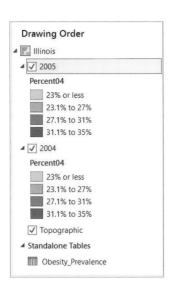

If you simply change the symbol field to show 2005 percentage values, the breaks you defined earlier will not be maintained. Instead, you will import the symbology scheme from the 2004 layer but make it display the Percent05 values.

3. With 2005 selected in the Contents pane, go to the Symbology pane, click the Options button, and click Import Symbology.

The Apply Symbology From Layer tool opens in the Geoprocessing pane.

4. Keep 2005 as the input layer. For Symbology Layer, choose 2004. Ensure that the source field points to Obesity_Prevalence.Percent04. Change the target field to Obesity_Prevalence.Percent05.

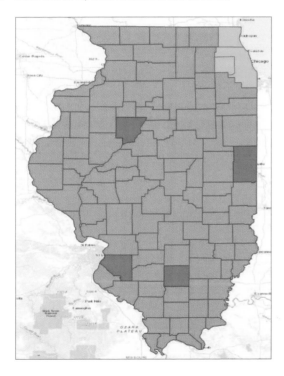

5. Run the tool, and examine the results.

6. Turn the 2005 layer on and off to see the difference between years.

In terms of map colors, in the 2005 map, two yellow counties have now advanced to light orange, and three light-orange counties have advanced to dark orange.

7. Continue steps 1 through 6 five more times: paste the 2004 layer, rename it (the next one will be **2006** and then **2007** and so on), import the 2004 symbology scheme, and change the target field value to the appropriate PercentXX value for the input layer.

Make sure the input layer is the current layer you are working with and the symbology layer is set to 2004.

8. Turn off all the obesity layers except 2004. One by one, turn on each subsequent year layer.

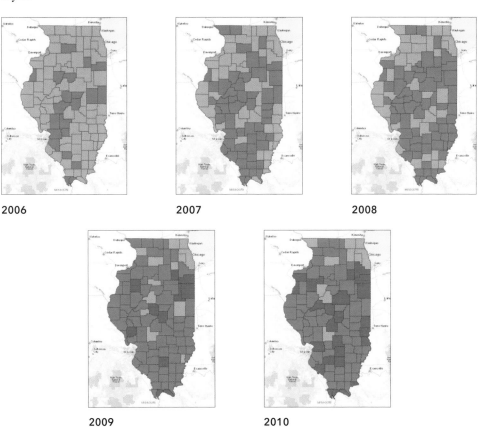

2006   2007   2008

2009   2010

The visual impact of the data is striking. With a few exceptions, almost every county's obesity prevalence goes up for each year.

Use the Swipe function to compare layers

1. Turn off and collapse all layers except 2004 and 2010.

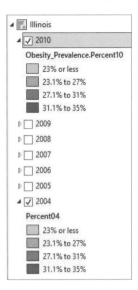

2. Use the Swipe function to reveal the change over six years: with the 2010 layer selected in the Contents pane, click the Feature Layer contextual tab, and then activate the Swipe tool.

3. Click the top of the map, and drag the arrow down to the bottom of the map. Swipe horizontally to observe the same effect.

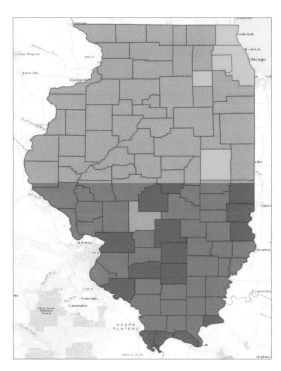

Overlay additional data

Some studies suggest that people with lower income have higher incidence of obesity and diabetes. You want to see if there is an obvious correlation between median income and obesity rates in Illinois counties.

1. In the Catalog pane, add IL_med_income.lyrx from the HealthStudy\Data folder.

This is a *layer file*, which is a saved symbology scheme that points to a specific source dataset. In this case, the source dataset is a shapefile by the same name, stored in the same folder—IL_med_income.shp. You can create a layer file from any symbolized layer in ArcGIS Pro: right-click the layer in the Contents pane, click Sharing, and click Save As Layer File. If you share a layer file, you must also share the source dataset along with it. You can create a *layer package*, which bundles the layer file along with the source data.

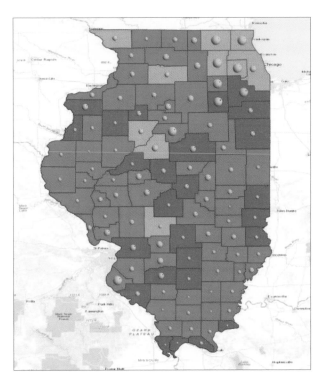

*Do you see a clear correlation between income
and 2010 obesity rates? Explain.*

Although you cannot draw scientific conclusions from a visual observation, you can determine whether further analysis is warranted.

ON YOUR OWN

Create a layer file for the 2004 and 2010 layers. Save them to the project folder.

2. Click the Explore tool to deactivate the Swipe tool, and save the project.

You are finished with this exercise. You will continue working with the same project for the
next exercise.

Exercise 3c: Calculate data statistics

Estimated time to complete: 25 minutes

Your project is filling out. Now you can easily present maps that show obesity prevalence rates per county, with a different map for seven consecutive years. Your committee has requested a map that shows the percentage change over the seven-year study period. You do not have this information, but you can create it. In this exercise, you will add a new attribute field and then populate the field with values. You will also calculate summary statistics for the state.

Exercise workflow

- Add a new field to the Illinois counties layer that shows the percentage change between two years (2004 to 2010).
- Use the Field Calculator tool to populate the new field.
- Display the new attributes on the map and modify the default labels.
- Use the Summary Statistics tool to create a table that shows the total number and average county percentage rate of obesity cases in Illinois as reported for the year 2010.
- Explore and configure the Infographics tool to get supplemental information about the study area.

Add a new field

To display the percentage change between layers, you need a percentage change field in your attribute table. You will create that field now, and then you will populate the field values all at once using the Field Calculator function.

1. Continue working in the HealthStudy project you have been working on in the previous two exercises.

   If you did not complete the previous exercises successfully, you may use HealthStudy2.aprx, found in the GTKAGPro\HealthStudy\Results folder.

   Rather than add another layer to your already busy map, you will create a new map to display the percentage change values.

2. In the Catalog pane, right-click Maps and click New Map.

   A new topographic map is added to the project. Switch between the maps by clicking the tabs at the top of the map display.

3. Name the new map **Illinois2**.

To rename, either click the word Map twice, slowly, to make it editable, or double-click to open the properties and change the name on the General tab.

4. Click the Illinois tab to switch back to the original map.

5. In the Contents pane, right-click 2010 and click Copy.

6. Switch to the Illinois2 map, and in the Contents pane, right-click Illinois2 and click Paste.

7. Change the new layer name to **Percent change**.

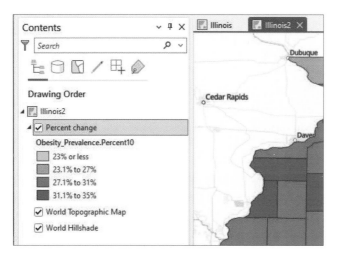

The layer name does not match what the layer is showing right now, but you will fix that issue soon.

8. In the Geoprocessing pane, find and open the Add Field tool, and complete these steps:
- For Input Table, click Percent Change.
- For Field Name, type **Perc_change**.
- For Field Type, click Double (64-bit floating point).

9. Maintain the other default settings, and run the tool.

As an alternative to searching in the Geoprocessing pane, you can open the Add Field tool from the geoprocessing tool gallery. On the Analysis tab, click through the gallery of tools until you find the Add Field tool under Manage Data. Click to open the tool. (If your application window is not maximized, you will see an Analysis Gallery button. Click it to access the gallery.)

You can customize the gallery–tools can be added, removed, or reordered. (Removing a tool from the gallery does not delete it from the system–you can always find any tool using the Search function.) If your gallery is customized, it will look different.

Input Table

Percent change

Field Name

Perc_change

Field Type

Double (64-bit floating point)

10. When the process is complete, open the table for the Percent change layer, and find the new field.

It is the last field in the table. It currently has no values, represented by <NULL>.

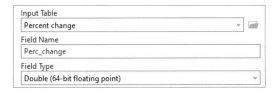

Number10	Percent10	Perc_change
2179	30	<Null>
7126	32.4	<Null>
12903	27.9	<Null>
1768	31.1	<Null>
10601	28.5	<Null>

You have added a permanent field to the source data of the Perc_change layer. Recall from exercise 3b that all the obesity layers in this map come from the same base data (the Illinois_cnty feature class). They look different because their symbology schemes point to different attribute fields. If you open the attribute tables for the other obesity prevalence layers (2010, 2009, 2008, and so on), you will see all the same attributes, including the new Perc_change field.

Calculate field values

1. In the attribute table, right-click the new field and click Calculate Field to open the Calculate Field tool.

The Calculate Field tool opens. You want the new field to show the percentage change between 2004 and 2010. Because the percentage values already exist, your calculation is a matter of simple subtraction. In the Calculate Field tool, the Input Table option should already point to Percent change.

2. For Field Name, click Perc_change.

3. Under Fields, double-click Percent10.

4. Click the subtraction symbol and double-click Percent04.
   Notice that the expression box fills in.

5. Click OK to run and close the tool, and then examine the new values in the table.

Number10	Percent10	Perc_change
2179	30	5.8
7126	32.4	7.9
12903	27.9	4.7
1768	31.1	6.2
10601	28.5	5.4

6. Close the attribute table.

Display a new field

Before the next step, you may need to save, close, and reopen the project.

1. Open the Symbology pane for the Percent change layer. Change Field to Perc_change.

2. At the top of the pane, click the Advanced Symbology Options button.

3. Expand the Format labels category, and change the number of decimal places to 1.

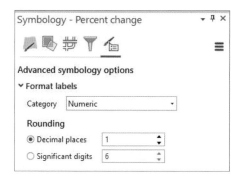

4. Click the Primary symbology button to return to the main Symbology pane.

5. Change the Color Scheme to Blues (4 Classes).

A ramp, which progresses from light to dark colors, will communicate increasing percentage change most clearly.

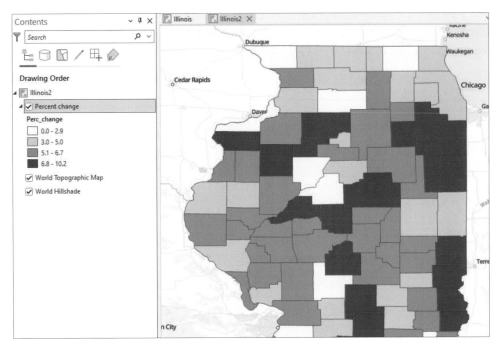

The Percent change layer shows which counties had the greatest increases between 2004 and 2010. These darker counties might be of interest to researchers and policymakers. The reason for the sudden changes should be investigated.

Calculate summary statistics

Finally, you want to present a table that shows the most pertinent statistics at a glance.

1. On the Analysis tab, expand the tool gallery to find and open the Summary Statistics tool. (Alternatively, search for Summary Statistics in the Geoprocessing pane.)

2. For Input Table, click the Percent Change layer. For Output Table, maintain the default name (Percentchange_Statistics).

3. For Field, click the down arrow and click Number10. For Statistic Type, maintain the default setting (Sum).

4. In the second row, click Percent10. For Statistic Type, click Mean.

The statistics that ArcGIS Pro generates will tell you the total number of obesity cases in Illinois in 2010, as well as the per-county average. You can also calculate the same statistics for the other years represented in your project.

5. Leave the optional Case Field parameter blank, and then run the tool.

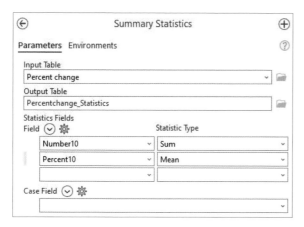

The output table is added to the bottom of the Contents pane.

6. Open and examine the table. Close it when you are finished.

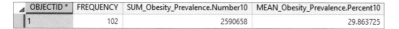

OBJECTID *	FREQUENCY	SUM_Obesity_Prevalence.Number10	MEAN_Obesity_Prevalence.Percent10
1	102	2590658	29.863725

You can now present numbers for the entire state. As of 2010, Illinois had 2,590,658 obesity cases.

Examine infographics

1. Go back to the Illinois map. On the Map tab, click the Infographics tool to activate Infographics, and click Alexander County at the southern tip of Illinois (see figure).

   Press the letter C on the keyboard to pan the map to the south, and scroll the mouse wheel to zoom in.

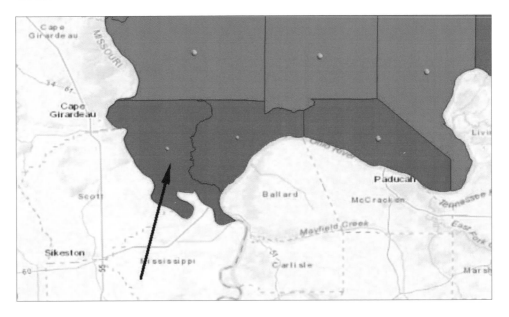

The Infographics tool is enabled only if you are signed in to your ArcGIS Online organizational account.

The Infographics window shows you information about a feature that is not stored in the dataset; rather, it is geoenrichment data that is served from ArcGIS Online. The information is presented as statistics, charts, or graphs.

2. Change the View to Slide Mode, advance the slides to see 2023 Households by Income (Esri), and compare Alexander County with the entire state of Illinois.

2023 Households by income (Esri)

The largest group: <$15,000 (18.6%)

The smallest group: $200,000+ (1.6%)

Indicator ▲	Value	Diff
<$15,000	18.6%	+9.0%
$15,000 - $24,999	14.8%	+7.9%
$25,000 - $34,999	12.2%	+5.2%
$35,000 - $49,999	16.5%	+6.1%
$50,000 - $74,999	17.6%	+1.5%
$75,000 - $99,999	9.9%	-3.0%
$100,000 - $149,999	7.0%	-10.5%
$150,000 - $199,999	1.8%	-7.0%
$200,000+	1.6%	-9.2%

Bars show deviation from   Illinois ▼

3. Click the arrows on the sides of the window to browse through the available data.

What percentage of households has an income of less than $15,000 per year?

4. Close the Infographics window when you have finished. Save the project.

You have done a great deal of research, discovery, and calculations for the Illinois counties health study project using ArcGIS Pro. Continue using this project for the next exercise.

Exercise 3d: Connect spatial datasets

Estimated time to complete: 15 minutes

In this short exercise, you will test one more hypothesis: that obesity rates correlate with higher numbers of food deserts, or low-income areas with limited access to full-service grocery stores.

Exercise workflow

- Create a relationship between a food desert layer (representing areas that are low income and have limited access to grocery stores) and the Illinois counties layer.
- Spatially join the layers so that you can generate a count of food deserts within each county.

Relate tables

Next, you will *relate* two tables, which allows you to quickly see relationships between separate datasets.

1. In ArcGIS Pro, continue working with the HealthStudy project, and activate the Illinois map. Turn off the IL_med_income and 2004 layer, confirm that all layers except for the 2010 layer are now off, and collapse their legends. Zoom to the 2010 layer.

    If you did not successfully complete the previous exercises, use HealthStudy2 .aprx in the HealthStudy\Results folder.

2. In the Catalog pane, from the HealthStudy\Data folder, add the IL_food_deserts.lyrx layer file. Drag it above the 2010 layer. Open its table and examine its attributes.

    This layer shows Illinois data for the census tract level rather than the county level. Notice the field named LILATracts. This field tells you if a census tract is considered a food desert (LILA stands for low-income, low-access). A value of 1 means that the tract is a food desert; a value of 0 means it is not. The symbology scheme is set to show only the tracts that have a LILATracts value of 1. Notice, too, that only rows that have a LILATracts value of 1 have a County value.

You will relate the IL_food_deserts table to the 2010 layer. That way, for each county, if there are statistics for food deserts, you can easily see them. You will relate the tables based on a common field: the County name field. Because only tracts that qualify as food deserts have a County name value, the tracts that are not considered food deserts are left out of the operation.

3. Close the table.

4. In the Contents pane, right-click 2010, point to Joins and Relates, and click Add Relate. Enter the following tool parameters, and then run the tool:
    - Layer Name or Table View: 2010
    - Input Relate Field: NAME
    - Relate Table: IL_food_deserts
    - Output Relate Field: County
    - Relate Name: **Relate1**
    - Cardinality: One to many

You are choosing a cardinality rule of one to many because each county may have more than one food desert.

5. On the Map tab, activate the Select tool, and click Knox County, as shown in the figure.

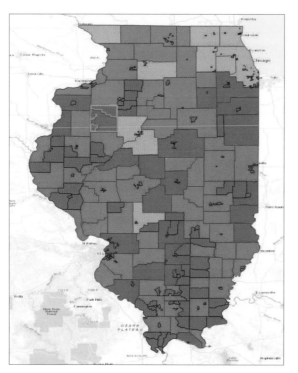

6. Open the 2010 attribute table and click the Show Selected Records button at the bottom left of the attribute table.

7. At the top right of the table, click the options menu, and click Related Data > IL_food_deserts.

The IL_food_deserts table opens, showing only the records for food desert tracts that share a name with the selected record in the other table.

When multiple tables are open, you can reposition them, float them, or stack them.

How many food deserts are in Knox County?

ON YOUR OWN

Select another county that appears to have multiple food deserts, and display the related records.

8. Close the tables, and clear any selections.

Spatially join data

Sometimes, rather than joining data based on a common attribute, you will want to join data based on location. This operation is called a *spatial join*. It allows you to define a spatial relationship between two layers (a target layer and a join layer) and combine their attributes in a new output layer. Depending on the types of data you want to join, you can choose how the spatial relationship is calculated. For instance, if you have county-level data but not state level, you can summarize county births for a state total, or you can average county income for a statewide average. You may also choose to generate a Count field that summarizes the number of features in the joined layer that *intersect* each feature in the target layer. Or you may generate a distance measurement between features in two layers. For example, you can define a relationship between city points and earthquake points based on distance, generating an output that shows the nearest city to each earthquake.

Here, you will spatially join the 2010 layer with the IL_food_deserts layer. The output will be a new geodatabase feature class, with the food desert count and affected population fields summarized for each county. Recall that LILATracts (low-income, low-access tracts) field values are either 0 or 1, with 1 indicating a food desert. By summarizing this field, you can quickly see how many food deserts are in each county. The POP_2010_1 field is the number of people who live in low-access areas. The field summary will show the total population of each county that may be affected by limited access to grocery stores.

1. Turn off the IL_food_deserts layer, and clear any selected features.

2. On the Analysis tab, in the Tools gallery, click Spatial Join (from the Overlay Data toolset).

3. Fill out the parameters as follows:
   - Target Features: 2010
   - Join Features: IL_food_deserts
   - Output Feature Class: **IL_food_deserts_SpatialJoin**
   - Join Operation: Join one to one
   - Keep All Target Features: checked
   - Match Option: Contains

4. Expand Fields. Under Field Map, scroll down and click LILATracts. For LILATracts, point to Edit Field Properties (pencil icon) and click to open the Field Properties window.

5. Under Table, select IL_food_deserts (1). On the right, click the down arrow and click Sum. Click OK to close the Field Properties window.

6. In the Spatial Join pane, under Field Map, scroll down and click POP2010_1. Point to Edit Field Properties and click to open the Field Properties window.

7. Under Table, select IL_food_deserts (1). On the right, click the down arrow and click Sum. Click OK to close the Field Properties window.

Output Fields lists all attributes available from both target and join features. You can add or remove fields or change their merge rules or properties.

8. Run the tool.

   When the operation is complete, a new layer is added to the map.

9. Symbolize the new layer using a single symbol: No Background:
   - In the Contents pane, right-click IL_food_deserts_SpatialJoin and click Symbology.
   - In the Symbology pane, under Primary Symbology, click Single Symbol (if not already chosen). Click the symbol color, scroll down toward the bottom of the ArcGIS 2D section, and click No Background.

Next, you will label the layer based on the LILATracts field.

10. With the IL_food_deserts_SpatialJoin layer selected in the Contents pane, click the Labeling contextual tab. In the Label Class group, keep Class set to Class 1, and set Field to LILATracts.

11. In the Contents pane, right-click the layer and click Label to turn the labels on.

Alternatively, click the Label button on the ribbon.

The labels show how many food deserts are in each county.

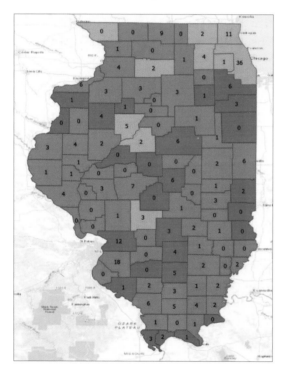

Although some counties have both high obesity prevalence and a high number of food desert tracts, there does not seem to be an obvious correlation at first glance. Some red counties have zero or only one food desert tract, while others have several.

Although you hoped to see some obvious correlative patterns between obesity, income, and food deserts in your initial visual analysis, the comparison of thematic maps is not the end of the story. Remember, you are mapping complicated problems that rarely have simple answers. Your next step would be to delve into more rigorous analysis techniques, using tools that model regression analysis to better compare your data variables. These advanced analysis methods are beyond the scope of this book, but if you are interested, refer to the ArcGIS Pro Tool Reference topic: Geoprocessing Tools > Spatial Statistics Toolbox > Modeling Spatial Relationships.

Even if you find no clear statistical relationship between variables, you can either explore other possible variables or recommend that the committee not limit its target audience to people with a lower income or even to people who live in a food desert. The committee might discuss advocating a large-scale education program delivered through various media to target individuals from all areas and walks of life.

12. Save and close the project.

Summary

This chapter is all about data relationships. You selected features by their spatial relationship to other features and then exported the selection to a new dataset. You turned nonspatial data into spatial data using a join operation that depended on an attribute relationship—a common field. You explored relationships within a single dataset when you calculated new fields and ran summary statistics. And you linked spatial datasets by relating their tables and performing a spatial join.

You should be getting comfortable using GIS to study problems and answer questions. In chapter 4, you will learn how to create and edit points, lines, and polygons.

Glossary terms

attribute query

attribute join

join table

input table

data type

graduated colors

manual interval classification

equal interval classification

defined interval classification

quantile classification

natural breaks (Jenks) classification

geometric interval classification

standard deviation classification

layer file

layer package

relate

spatial join

intersect

## CHAPTER 4

Creating and editing spatial data

Exercise objectives

4a: Build a geodatabase
- Convert shapefiles to geodatabase feature classes.
- Map x,y points.
- Establish an attribute domain.

4b: Create features
- Configure snapping options.
- Create a line feature.
- Enter attribute data.

4c: Modify features
- Split polygons.
- Merge polygons.
- Modify lines and points.

You are probably getting comfortable exploring, querying, combining, and displaying spatial data. But what if you want to create it from scratch? This chapter teaches you how to create and edit spatial data.

Scenario: You work in the GIS department of Troutdale, Oregon. One of your jobs is to maintain the city's GIS database. You are considering converting all the city's shapefiles into the geodatabase feature class format because you want to take advantage of the benefits offered by the geodatabase. You will begin by converting just a few of the city's datasets so that you can decide whether it makes sense to convert the city's extensive data library. Then you will resume your maintenance duties as you create some new features and modify others.

WHY USE A GEODATABASE?

ArcGIS spatial data formats include raster data, such as surfaces, terrains, and imagery, and vector data. Tabular data that records x,y locations and other attributes is also a type of spatial data that you can integrate into ArcGIS Pro, as you will see in the upcoming exercise. The most common data formats are the shapefile and geodatabase feature class. Both types of data comprise points, lines, or polygons that represent geographic objects of the same kind, such as countries or rivers.

A *shapefile* is a simple, stand-alone data format. It stores geometry and attribute data for one set of features. FireHydrants.shp, for example, might represent the point locations of fire hydrants. Shapefiles are composed of several files that have different extensions, although you will see only one item listed in the Catalog pane.

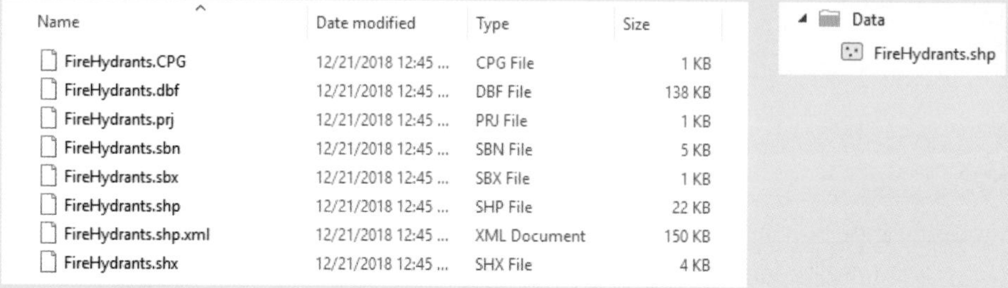

Name	Date modified	Type	Size
FireHydrants.CPG	12/21/2018 12:45 ...	CPG File	1 KB
FireHydrants.dbf	12/21/2018 12:45 ...	DBF File	138 KB
FireHydrants.prj	12/21/2018 12:45 ...	PRJ File	1 KB
FireHydrants.sbn	12/21/2018 12:45 ...	SBN File	5 KB
FireHydrants.sbx	12/21/2018 12:45 ...	SBX File	1 KB
FireHydrants.shp	12/21/2018 12:45 ...	SHP File	22 KB
FireHydrants.shp.xml	12/21/2018 12:45 ...	XML Document	150 KB
FireHydrants.shx	12/21/2018 12:45 ...	SHX File	4 KB

▲ 📁 Data
  ⊡ FireHydrants.shp

Figure 4.1. A shapefile's system files, and how it looks in ArcGIS Pro.

A *geodatabase*, by contrast, is a storage container, in which sets of features are grouped into *feature classes*. The geodatabase can also store rasters and special geodatabase elements that facilitate capabilities that are not available with other data formats. For example, World.gdb might contain a polygon feature class of world countries, a line feature class of world rivers, a point feature class of world cities, and so on.

When working with ArcGIS Pro, you can add data from a variety of raster, vector, and tabular formats. When you add spatial data to ArcGIS Pro, you see it on the map, work with it as a layer, and usually are not interested in the format of the source file.

One of the advantages of a geodatabase is the ability to store multiple datasets, including rasters, which makes the geodatabase a great option for organizing data. You can group feature classes into a *feature dataset*—for example, sewer mains (lines) and manhole locations (points) can be grouped together in a sewer feature dataset.

When shapefiles are imported into a geodatabase, the size on disk is dramatically reduced, so the geodatabase is an ideal model for data sharing.

Another advantage of using a geodatabase is the ability to create an attribute domain, which establishes and enforces valid values or ranges of values for an attribute field and minimizes data entry mistakes by prohibiting invalid values. For example, if Open and

Closed are the domain values for a water valve status field, you cannot enter intthe attribute table any values other than Open or Closed.

You can also create subtypes within geodatabase feature classes. A typical example is streets: for example, you can create subtypes for local, arterial, and collector streets. You can assign different default attributes or attribute domains to each; for example, local streets might have a default speed limit of 25 mph, whereas arterial streets might have a default speed limit of 35 mph. Or you can apply a different speed limit domain to each subtype.

Finally, the geodatabase can store behavior rules. Thus, a geodatabase can support specialized types of data, such as annotation feature classes, cartographic representations, geodatabase topologies, and networks, which all have specific behaviors. These special data types are increasingly important to GIS, although they are beyond the scope of this book.

DATA

In the C:\GTKAGPro\CityMaintain\Data folder:

- FireHydrants.shp—a point shapefile that represents fire hydrants in Troutdale *(Source: City of Troutdale)*
- WaterLines.shp—a line shapefile that represents water mains in Troutdale *(Source: City of Troutdale)*
- WaterPressureZones.shp—a polygon shapefile that represents water pressure zones in Troutdale *(Source: City of Troutdale)*
- Wells.shp—a point shapefile that represents wells in Troutdale *(Source: City of Troutdale)*
- Valves.csv—a table that contains x,y coordinate values for the locations of water valves in Troutdale *(Source: City of Troutdale)*

Exercise 4a: Build a geodatabase

Estimated time to complete: 30 minutes

In this exercise, you will test the feasibility of switching the city's data collection to the geodatabase model. You will populate an empty geodatabase by converting shapefiles and mapping x,y location values found in a nonspatial table. All the outputs will be placed in the geodatabase. Then you will create an attribute domain in the geodatabase and apply it to the WaterLines feature class. This will ensure that valid values are entered into the table during the editing process.

Exercise workflow

- Convert shapefiles that represent various city water features into geodatabase feature classes.
- Convert a table that stores attributes and x,y location points (but no Shape field) into a point feature class.
- In the CityMaintain geodatabase, create a coded domain for the pipe diameter field so that entries are constrained to the standard pipe diameter measurements of four, six, eight, or twelve inches.
- Apply the domain to the WaterLines feature class.

Convert shapefiles to geodatabase feature classes

1. In ArcGIS Pro, open CityMaintain.aprx, found in your C:\GTKAGPro\CityMaintain folder. At this point, the map contains only a basemap layer (Streets).

2. In the Catalog pane, expand Folders > CityMaintain > Data.

The project includes an empty geodatabase (CityMaintain.gdb), an empty toolbox (City-Maintain.tbx), and a data folder that has four shapefiles (.shp) and a table (.csv). You want to convert these shapefiles into geodatabase feature classes in CityMaintain.gdb.

Suppose you want to create a geodatabase. To do so, right-click Databases in the Catalog pane, and click New File Geodatabase. Notice that you can also add an existing file geodatabase or create a connection to an enterprise geodatabase that resides on a server.

3. Open the Geoprocessing pane.

REMIND ME HOW

ArcGIS Pro remembers the panes that you have opened before, so the Geoprocessing pane may already be open. If it is in the same space as the Catalog pane, click the Geoprocessing tab. Otherwise, on the ArcGIS Pro ribbon, click the Analysis tab, and click the Tools button in the Geoprocessing group.

You will search for a tool to help you convert the shapefiles into geodatabase feature classes.

4. In the Search box of the Geoprocessing pane, type **Feature Class To Geodatabase**, and in the search results, read the ToolTip for the Feature Class To Geodatabase tool (in the Conversion Tools group).

> **Feature Class To Geodatabase** (Conversion Tools)
> Converts one or more feature classes or feature layers to geodatabase feature classes.

This tool, a script tool, is the one you want.

In the Geoprocessing pane, built-in tools are denoted with a hammer icon, and script tools are denoted with a scroll icon.

5. Open the Feature Class To Geodatabase tool. For Input Features, click the browse button (the folder icon) to open the Input Features window.

6. Browse to C:\GTKAGPro\CityMaintain\Data, click FireHydrants.shp, press Ctrl on your keyboard, and click the other three shapefiles (WaterLines.shp, WaterPressureZones. shp, and Wells.shp).

7. Click OK to load all four shapefiles into the tool as input features.

You can also drag the files from the Catalog pane into the tool.

8. For Output Geodatabase, click the browse button to open the Output Geodatabase window, browse to CityMaintain.gdb, and select it. Click OK.

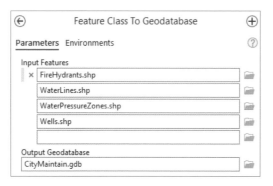

9. Run the tool. When the process is complete, keep the Geoprocessing pane open.

10. In the Catalog pane, expand the CityMaintain geodatabase to see the four new feature classes. If necessary, right-click the geodatabase and click Refresh to refresh the view.

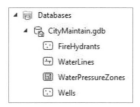

In the Catalog pane, you can see the CityMaintain geodatabase under Folders > CityMaintain.

Notice that when you use the script tool, the feature classes are not automatically added to the map.

11. Drag the new geodatabase feature classes to the map. (Hint: Use Shift or Ctrl to select them all at once.)

12. Change the WaterLines color by right-clicking its symbol in the Contents pane and choosing a dark blue (such as Lapis Lazuli) from the color palette.

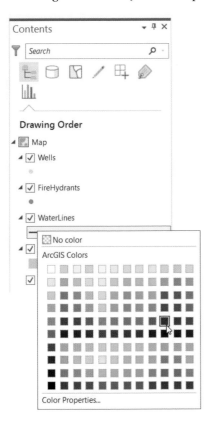

13. Change the symbology of the other layers as indicated. To do so, click the current symbol under the layer in the Contents pane, which opens the Symbology pane. For each layer, choose the following symbol from the symbol gallery, using the search function as needed:

- Wells: Pentagon 1
- FireHydrants: Diamond 3
- WaterPressureZones: Black Outline (1pt)

Point to the symbol to see the entire name, including line thickness.

14. Zoom to the Wells layer.

REMIND ME HOW

Right-click Wells in the Contents pane and click Zoom To Layer.

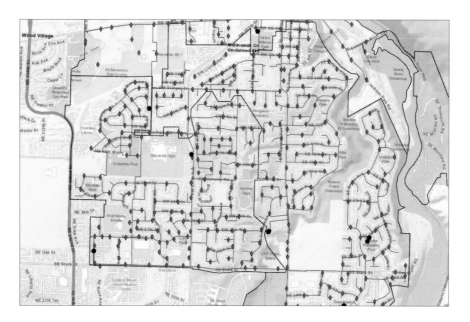

15. Save the project.

It is a good habit to save often.

Map x,y points

You also have a stand-alone table in your project that you want to convert to a feature class.

1. Drag the Valves.csv table from the CityMaintain\Data folder to the map and open it.

	Updated_Date	Valve_Type	Valve_Size	Valve_Useage	Cover	Lid_Type	Valve_Cond
1	8/12/2014	GV	2	Blow Off	Asphalt	910	Good
2	11/22/2011	GV	6	Hydrant	Asphalt	910	Good
3	3/28/2012	GV	8	Inline	Asphalt	910	Good
4	8/12/2014	GV	4	Inline	Asphalt	910	Good
5	3/28/2012	GV	2	Blow Off	Asphalt	910	Good

This table is nonspatial—that is, it does not have a well-defined geometry as feature classes do (there is no Shape field), so it cannot be displayed on a map. But the table contains spatial attributes in the form of x,y coordinates, which means that you can easily turn it into mappable data.

2. Scroll all the way to the right side of the table to see the POINT_X and POINT_Y fields. Close the table when you are finished.

3. In the Geoprocessing pane, search for the XY Table To Point tool (Data Management), and read the description or ToolTip.

COORDINATE SYSTEMS

All spatial data has a *coordinate system*, an important component of its spatial reference. The coordinate system defines features' positions on the earth's surface. Coordinate systems can be geographic or projected.

A *geographic coordinate system* uses latitude and longitude to define the locations of points on the surface of a sphere or *spheroid*. A *projected coordinate system* uses a mathematical equation (the *map projection*) to transform latitude and longitude coordinates into Cartesian or planar coordinates for display on a flat map. Particularly for detailed analysis or editing, it is important for all your layers to be in the same projected coordinate system.

ArcGIS employs *on-the-fly projection*, which means that it applies the projected coordinate system of the first layer added to all subsequent layers.

The following images show a world map in the WGS84 geographic coordinate system on the left and the Winkel-Tripel projected coordinate system on the right.

For more information, see the ArcGIS Pro Help topic: Maps and Scenes > Map and Scene Properties > Coordinate Systems > Coordinate Systems, Projections, and Transformations.

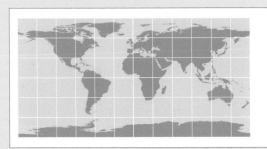 

You will create a point feature class using the x,y coordinates stored in the table. You will make sure it is properly placed on the map by importing the projected coordinate system (see sidebar) of one of the other city layers.

4. Open the XY Table To Point tool, and change these parameters (some of them may autopopulate):
   - For Input Table, click Valves.csv.
   - For Output Feature Class, keep the default of Valves_XYTableToPoint.
   - For X Field, keep the default of POINT_X.
   - For Y Field, keep the default of POINT_Y.
   - Leave Z Field blank.
   - For Coordinate System, click Wells.

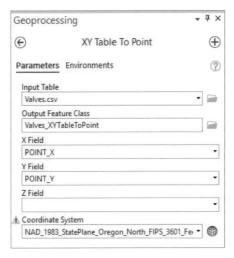

To increase your understanding, point to the information icons to read the tip for each tool parameter.

5. Notice the optional warning, and run the tool.

   Don't worry about the warnings; this dataset does not have z-values (which record elevation or depth).

   The feature class is added to the project geodatabase and to the map. (You may need to refresh the geodatabase in the Catalog pane to see it.)

6. In the Contents pane, remove Valves.csv (the stand-alone table), and change the new layer name to **Valves**.

7. Change the Valves symbol to Circle 4, in Moorea Blue.

REMIND ME HOW

In the Contents pane, click the default symbol for Valves. In the Symbology pane symbol gallery, click Circle 4 and click Properties. Change the color chip to Moorea Blue (four rows from the top), and click Apply.

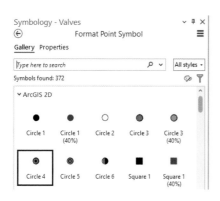
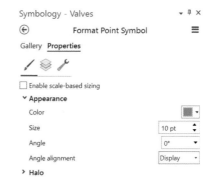

METADATA

Whenever you create new data, it is best to create *metadata*—textual information about the dataset—especially before sharing. To do so, right-click the dataset in the Catalog pane, and click View Metadata. A Catalog view appears in the display area, and new options are available on the ribbon. On the Catalog tab, in the Metadata group, click the Edit tool. Enter your tags (which facilitate data searches), a summary of the dataset, and anything else you feel is necessary (such as data credits). When you are finished, click Save, and close the Catalog view.

Establish an attribute domain

To diminish the likelihood of data entry errors, you will set an *attribute domain* in your geodatabase—a set of valid values, or a numerical range, for the attributes in each field. Specifically, you will create a coded value domain for the pipe diameter field of the WaterLines feature class. Pipe diameters must be four, six, eight, or twelve inches.

1. Open the WaterLines attribute table. Click the Options button and then click Fields
View.

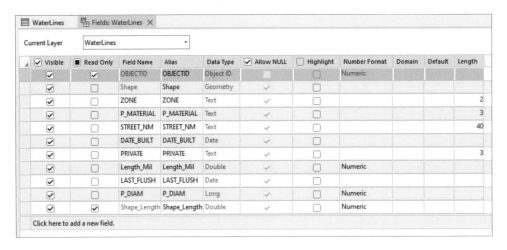

You now see the Fields tab, in which you can manage fields to make edits to a table's fields,
modify their properties, delete fields, or create new ones. A Fields tab also appears on the
main ribbon.

2. At the bottom of the table, find the P_DIAM field. Select the cell in the Domain column.
Click the cell again, and from the options, click Add New Coded Value Domain.

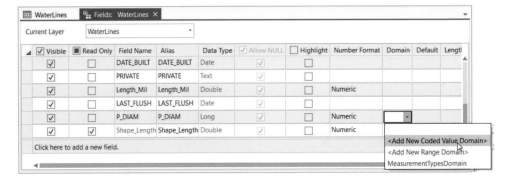

This step opens the Domains table. A Domains tab also appears on the main ribbon. You may notice a Measurement Types domain, created by default.

3. Under MeasurementTypesDomain in the Domain Name column, double-click to edit the field, and type **PipeDiam**. For Description, type **Pipe diameter in inches**.

4. Maintain the default Field Type (Long), Domain Type (Coded Value Domain), and Split and Merge Policies (Default).

5. Under Code, type **4**. Under Description, type **4**, and then press Enter to create a new row.

6. Create three more Code and Description rows; and type **6**, **6**; then **8**, **8**; and finally, **12**, **12**.

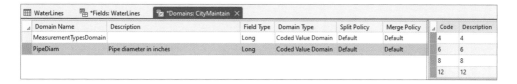

Domain Name	Description	Field Type	Domain Type	Split Policy	Merge Policy	Code	Description
MeasurementTypesDomain		Long	Coded Value Domain	Default	Default	4	4
PipeDiam	Pipe diameter in inches	Long	Coded Value Domain	Default	Default	6	6
						8	8
						12	12

7. On the Domains tab on the main ribbon, click Save.

   In this case, it makes sense to use the pipe diameter as both the code and the description. In other situations, the code and description may be different. For instance, say that you want to establish a domain to constrain the soil type attributes within a study area. You can assign an alphanumeric code to represent each soil type, which is spelled out in the description column.

   For more about attribute domains, go to the ArcGIS Pro Help topic: Data > Geodatabases > Data Design > "Introduction to Attribute Domains."

8. Click the tab for the Fields table. For the P_DIAM field, confirm that PipeDiam appears in the Domain column. In the next column on the right, under Default, click 4 in the drop-down list, and click Save on the Fields tab of the ribbon.

Visible	Read Only	Field Name	Alias	Data Type	Allow NULL	Highlight	Number Format	Domain	Default	Length
✓	✓	OBJECTID	OBJECTID	Object ID		☐	Numeric			
✓	☐	Shape	Shape	Geometry	✓	☐				
✓	☐	ZONE	ZONE	Text	✓	☐				2
✓	☐	P_MATERIAL	P_MATERIAL	Text	✓	☐				3
✓	☐	STREET_NM	STREET_NM	Text	✓	☐				40
✓	☐	DATE_BUILT	DATE_BUILT	Date	✓	☐				
✓	☐	PRIVATE	PRIVATE	Text	✓	☐				3
✓	☐	Length_Mil	Length_Mil	Double	✓	☐	Numeric			
✓	☐	LAST_FLUSH	LAST_FLUSH	Date	✓	☐				
✓	☐	P_DIAM	P_DIAM	Long	✓	☐	Numeric	PipeDiam	4	
✓	✓	Shape_Length	Shape_Length	Double	✓	☐	Numeric			

Now when you enter or edit waterline feature attributes, you will be constrained to enter pipe diameters of 4, 6, 8, or 12.

ON YOUR OWN

If you want, change the P_DIAM *alias* to **Pipe Diameter**.

9. Save changes to the Fields table. Close the tables, and save the project.
   You can continue to the next exercise, or exit ArcGIS Pro and come back later.

Exercise 4b: Create features

Estimated time to complete: 30 minutes

The WaterLines layer is missing a line feature. You will create it in this exercise, using a feature template, construction tools, and editing options.

Exercise workflow

- Configure the snapping options to facilitate editing.
- Create a new waterline that connects to existing water valve points.
- Enter attributes for the new waterline feature.

Configure snapping options

Snapping is an editing option that acts like a magnet. When editing, if the point you create is within a specified distance of another feature's *vertex*, *endpoints*, *edge*, or *intersection*, it will jump (or snap) to coincide with another feature. Snapping allows you to accurately connect features, such as waterlines and valves, without impossibly precise sketching. You can enable various snapping configurations—for example, you might choose to snap to the endpoint of a line or the vertex of a polygon, or both.

To better decide which snapping options will benefit you, you will display the vertices of a feature that is next to one that you will be creating.

1. Continue working in the CityMaintain project you worked on in exercise 4a.

   *If you did not successfully complete the previous exercise, open CityMaintain
   .aprx, add to the map all the feature classes found in CityMaintain\Results\
   CityMaintain2.gdb, and symbolize them as you want.*

   First, you will change the selection and editing status of layers so you can more easily select and edit the features you want.

2. In the Contents pane, click the List By Selection button. Turn off all layers except WaterLines.

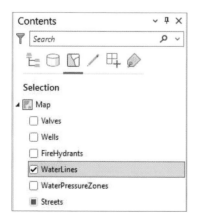

3. In the Contents pane, click the List By Editing button. Again, turn off all layers except WaterLines.

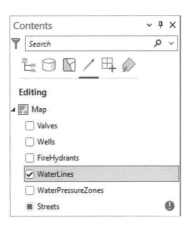

4. Return to the Drawing Order view of the Contents pane.

5. Go to the Water Lines bookmark on the Map tab.

Your dataset has two missing waterlines. Every valve should be connected to a waterline.

6. Click the Edit tab on the ribbon.
   Before you start creating, look at the existing line that runs along SW 19th Way.

7. To see the composition of the line, click the Select tool on the Edit tab, and then click the line, which is highlighted in cyan.

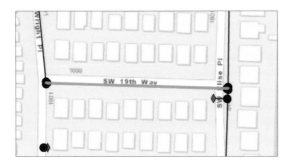

8. On the Edit tab on the ribbon, in the Tools group, click the Edit Vertices tool to display the vertices of the sketch.

The Modify Features pane opens and shows the x,y location of each vertex of the selected feature.

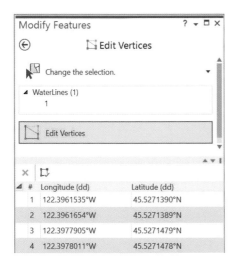

The Vertices toolbar appears on your map. It allows you to easily add or delete vertices and finish or cancel edits.

Refer to the Modify Features pane:
How many vertices does the selected line have?

9. Zoom to the corner of SW 19th Way and SW Wright Place to examine the selected line's vertices.

Here is some practical advice for detailed editing work: maximize your window, and zoom in closely. To pan or zoom the map without deactivating the current edit tool, use the mouse wheel shortcut for panning and zooming: scroll the wheel button to zoom, and press and drag the wheel to pan (or press C on your keyboard while dragging the map). Try it—it is helpful when editing data.

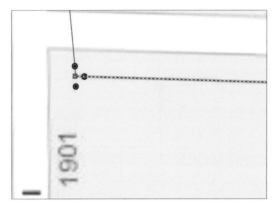

All lines have one vertex at each end and an unlimited number in between. By default, endpoints are red squares; other vertices are green squares. In this case, there is a vertex placed at the point at which the line intersects a valve and one at either end of the line. (Turn Valves off and on to see better.)

10. Clear the selection.

11. On the Edit tab, click the Snapping button to turn it on (when it is shaded in blue, snapping is on).

You can configure the Snapping tool so that it snaps only to certain places, such as points, line ends, or polygon edges. When a snapping option is turned on, its icon is highlighted in the configuration options.

12. Click the Snapping button down arrow, and configure the tool so that only point snapping and vertex snapping are turned on (again, blue shading indicates on).

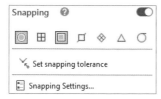

This configuration will allow you to snap to existing line vertices and valve points as you create your waterline features.

Create a line feature

You will create a new, L-shaped waterline that begins and ends at existing waterline junctions and intersects existing valves. The line you are about to create is shown selected (highlighted in cyan) in the figure.

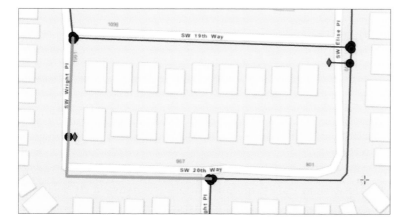

1. On the Edit tab, click the Create button.

The Create Features pane opens. The Create Features pane shows the available *feature templates* you can use to create features. The templates match the layer names in the Contents pane by default. In this case, only one template is available, because you marked only one layer—WaterLines—as editable in the Contents pane.

You can also choose which construction tools you want to show in the Create Features pane or turn on a prompt for entering one or more attributes when the feature is drawn. To configure feature templates, right-click the template name, and click Properties. You are not limited to a single symbol for each feature template. If you have a layer that has graduated or unique symbols for certain feature types, all the symbols will be available in the feature template. You choose the correct one based on the attributes of the feature you are creating.

For more information, go to the ArcGIS Pro Help topic: Data > Edit geographic data > Feature templates.

2. In the Create Features pane, click the WaterLines feature template, and make sure the Line tool is active.

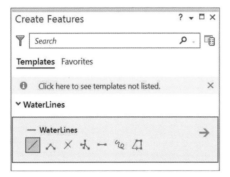

The Construction toolbar opens in the map view. This toolbar provides more tool options for creating line features.

3. Move your mouse pointer close to the junction of the two existing waterlines. Notice that when it gets close, a SnapTip (WaterLines: Vertex) appears.

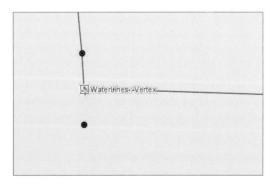

4. Zoom in closely. With the SnapTip visible, click to place a vertex at the junction of the existing waterlines.

Because of snapping, your click does not have to be so precise. If you see the SnapTip, the new vertex will be coincident with the existing vertex—that is, they are directly on top of each other.

5. Move your mouse to the valve south of the point you just clicked. When you see the Valves: Point SnapTip, click to place another vertex.

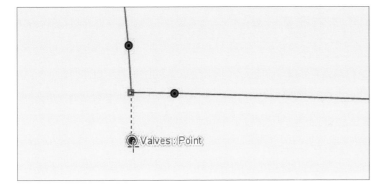

If you place a vertex incorrectly, click the Undo button at the top of the application.

6. Zoom out and pan south to the next valve. When you see the SnapTip, place another waterline vertex.

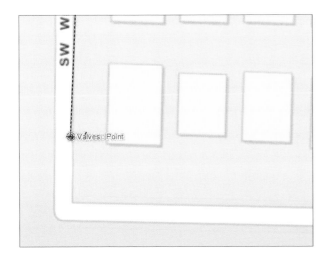

The next vertex you will place is where the waterline forms a right angle. There is no water valve to snap to, so you will use two edit commands to ensure proper placement: Distance and Deflection. You will constrain the next segment to a specific length of 58.5 feet, and you will ensure the straightness of the line by setting a deflection angle of 0 decimal degrees.

7. Right-click anywhere on the map. On the menu, click Distance. In the Distance pop-up window, type **58.5**, ensure the units are set to ft, and press Enter.

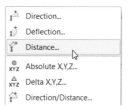

8. Right-click again. On the menu, click Deflection. In the Deflection pop-up window, enter an angle of **0** dd.

Your next line segment is now constrained to the exact distance you entered, with no deflection from the previous line. The vertex is placed for you.

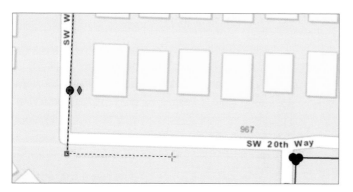

There are just two vertices left.

9. Extend the line parallel to SW 20th Way. Zoom in on the cluster of three valves. Snap to the first valve point, and click to insert a vertex.

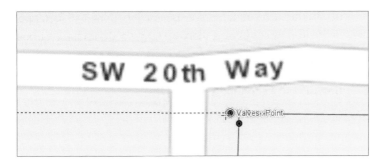

10. Snap to the WaterLines vertex to place your final vertex.

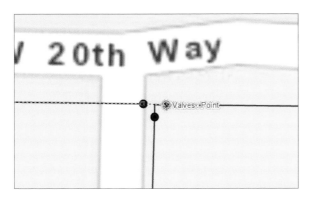

11. Move your pointer slightly away from the last vertex placed, right-click, and on the context menu, click Finish.

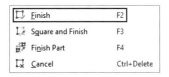

12. Zoom out (or go to the Water Lines bookmark on the Map tab) to see the completed new waterline. Click Clear to clear the selection so you can see the line.

ON YOUR OWN

Create the waterline between the valve and the fire hydrant on SW Wright Place. Try double-clicking to finish the sketch this time. When you are finished, clear the selection.

13. Save your edits (on the Edit tab, click Save). Click Yes to confirm.

Enter attribute data

Now you will enter attributes for the new waterlines.

1. Open the WaterLines attribute table. Locate the first feature you created (OBJECTID 2308) and show only the selected record.

2. Click twice inside the cell below the ZONE field to make it editable. Type **1**, and then press Tab to go to the next cell. In the P_MATERIAL field, type the capital letters **DI** (for ductile iron).

3. Enter these remaining attributes:
   - STREET_NM: **SW 20th Way/Wright Place**
   - DATE_BUILT: **3/3/2015, 12:00 a.m.**
   - PRIVATE: **N**
   - Length_Mil: **0.079449**
   - LAST_FLUSH: **4/15/2018, 12:00 a.m.**
   - P_DIAM: **8**

Recall that in exercise 4a, you created a coded domain rule for the P_DIAM field, and set the default value to 4. Notice that you are constrained to choose only a valid measurement: 4, 6, 8, or 12.

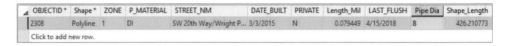

OBJECTID *	Shape *	ZONE	P_MATERIAL	STREET_NM	DATE_BUILT	PRIVATE	Length_Mil	LAST_FLUSH	Pipe Dia	Shape_Length
2308	Polyline	1	DI	SW 20th Way/Wright P...	3/3/2015	N	0.079449	4/15/2018	8	426.210773
Click to add new row.										

If you want to enter attributes before creating the feature, in the Create Features pane, click the blue arrow next to the appropriate feature template. Then enter attributes for the feature you are about to create.

You have successfully created a new line feature, complete with attribute data.

4. Save your edits, close the attribute table, and save the project.

You will continue working on the project in the next exercise.

Exercise 4c: Modify features

Estimated time to complete: 35 minutes

Sometimes spatial data contains mistakes that must be corrected, and sometimes feature locations or boundaries change. In this exercise, you will modify existing features in the City-Maintain geodatabase.

To stabilize water pressure throughout Troutdale, two new water towers were built, resulting in changes to water pressure zones. You will split an existing zone into two zones and then merge one of the new, smaller zones into an adjacent zone.

GIS IN THE WORLD: MUNICIPAL GIS, FROM DESKTOP SOFTWARE TO WEB APPS

Riverside County, California, has been incorporating GIS into its operations since 1989. Its GIS team has developed GIS-enabled solutions, such as tracking fire damage, mapping bat locations (and rabies testing results), and maintaining city and county data, including transportation and cadastral data. It now incorporates ArcGIS Online services into its business model to serve data and provide a space for sharing relevant maps and apps.

Read more about it in *ArcNews*, "Riverside County Takes GIS to the Next Level": www.esri.com/esri-news/arcnews/spring14articles/riverside-county-takes-gis-to-the-next-level.

Figure 4.3. The Riverside County Spatial Data portal. Using ArcGIS Online, you can find many map services and layers hosted by Riverside County, California.

Exercise workflow

- Split a water pressure zone into two separate features.
- Merge two water pressure zones into a single feature.
- Move an erroneously placed waterline.
- Permanently delete a well.

Split polygons

1. Continue working in the CityMaintain project.

   If you did not successfully complete exercises 4a and 4b, open CityMaintain .aprx, add all the feature classes from CityMaintain\Results\CityMaintain2.gdb, and symbolize them as you want.

2. Zoom to the WaterPressureZones layer.

3. In the Contents pane, in the List by Drawing Order view, turn off Valves, FireHydrants, and WaterLines.

    Only the Wells, WaterPressureZones, and Streets layers are visible.

4. In the List by Selection view in the Contents pane, make Wells, WaterLines, and WaterPressureZones the only selectable layers.

5. In the List by Editing view in the Contents pane, make the same three layers editable.

6. Return to the List by Drawing Order view in the Contents pane.

7. On the Map tab (or the Quick Access Toolbar, if you placed it there), click the Explore tool, and click the water pressure zone just south of Troutdale and west of the river (shown in the figure) to identify it. Take note of the OBJECTID, ZONE, and Shape_Area attributes.

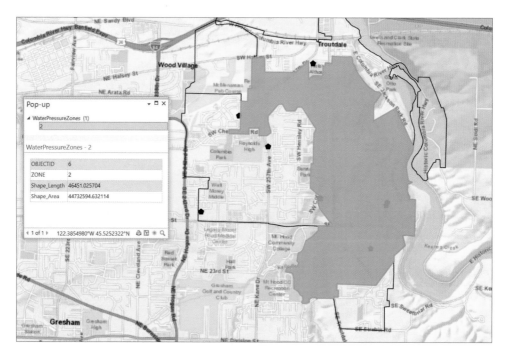

This is the feature you are going to split in two.

8. Close the pop-up window.

9. On the Edit tab, check your Snapping tool options to verify that point snapping and vertex snapping are turned on.

    If you have not made changes since the previous exercise, they should be on; ArcGIS Pro recalls your previous settings.

10. Using the Select tool, select water pressure zone 2, and click the Split tool in the Tools gallery.

    Alternatively, if necessary, click the Modify button to open the Modify Features pane, and select the Split tool from there.

11. Zooming in as necessary, place vertices in the approximate locations shown in the figure. Click to add a vertex outside the zone, add another vertex inside the zone, and add a vertex outside. Double-click to end the operation and complete the split.

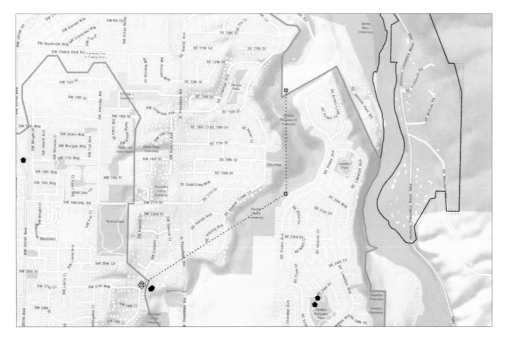

You can finish a sketch several ways: double-click, press F2 on your keyboard, right-click and click Finish, or click the Finish button on the pop-up toolbar.

One polygon feature has become two. You may notice when the split is complete that each new feature flashes briefly.

12. Use the Explore tool to identify each water pressure zone.

Notice the updated attributes. The northernmost zone has a new OBJECTID value but has maintained the ZONE attribute of 2.

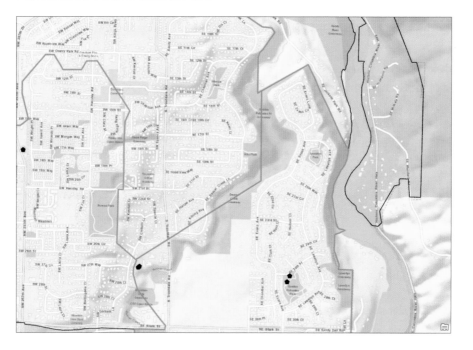

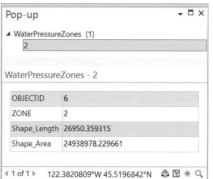

Pop-up			Pop-up		
▲ WaterPressureZones (1)			▲ WaterPressureZones (1)		
2			2		
WaterPressureZones - 2			WaterPressureZones - 2		
OBJECTID	7		OBJECTID	6	
ZONE	2		ZONE	2	
Shape_Length	29158.724709		Shape_Length	26950.359315	
Shape_Area	19793616.495357		Shape_Area	24938978.229661	
◀ 1 of 1 ▶ 122.3898043°W 45.5322223°N			◀ 1 of 1 ▶ 122.3820809°W 45.5196842°N		

Your Shape_Area values may be slightly different. Your OBJECTID values may also be different, depending on your editing history.

What happened to the Shape_Area value
of the original water pressure zone?

13. Clear the selection, and save your edits.

> **ON YOUR OWN**
>
> There are now two polygons identified as zone 2. Edit the WaterPressureZones attribute table, giving the northernmost of the new polygons a new ZONE value of **6**.

Merge polygons

Now you will merge one of the new water pressure zones (the southernmost result of the split operation) with the smaller one below it. The zones you will merge are indicated by the blue circle in the figure.

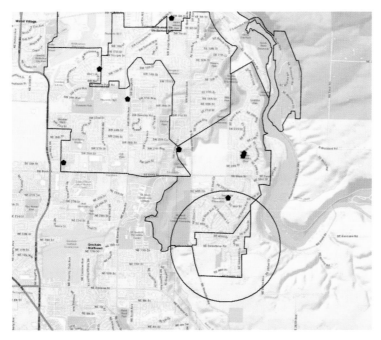

1. On the Edit tab, in the Tools gallery, click the Merge tool.

Another way to find edit tools: click the Modify button in the Features group, and click All Tools.

The Merge tool prompts you to select two or more features.

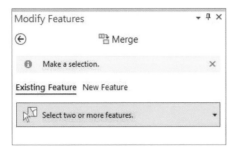

2. Zoom in to the polygons to be merged. Using the Select tool, click one of the polygons, and then press the Shift key while clicking the other polygon.

   Now both water pressure zones are selected.

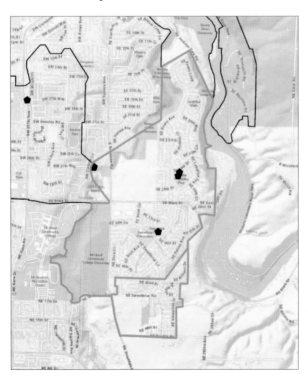

The Select tool can be found on both the Map and the Edit tabs.

The Modify Features pane is populated with information regarding the two selected zones. It tells you that you are about to merge zones 2 and 6.

3. In the Modify Features pane, under Features to Merge, choose to preserve the attributes of zone 2 by clicking 2 to highlight the zone on the map.

This step is a good visual confirmation of what you are about to do.

4. Click Merge to run the operation.

The map shows the two features merged into a single feature.

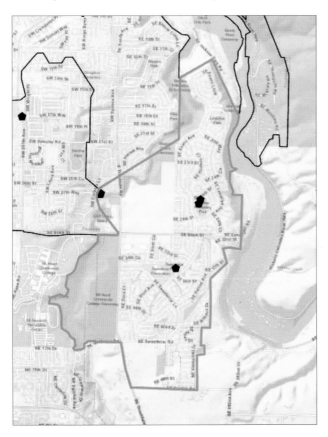

5. Using the Explore tool, display the attribute pop-up window for the new larger water pressure zone.

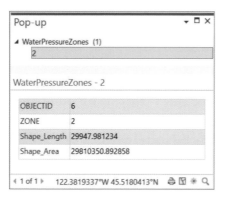

Notice that the new larger polygon has kept the ZONE value of 2.

6. Clear the selection, and close the pop-up window.

Modify lines and points

A waterline was placed incorrectly. You will fix the error by moving its vertices.

1. Turn on the WaterLines layer.

2. Go to the Move bookmark and reveal the vertices for the waterline shown in the figure.

REMIND ME HOW

On the Edit tab, activate the Select tool, click the line feature, and click the Edit Vertices tool.

The waterline should run alongside SE Pelton Avenue.

3. One at a time, place your mouse pointer over each vertex, as circled in the figure, and drag it to the approximate location indicated by the arrows along SE Pelton Avenue.

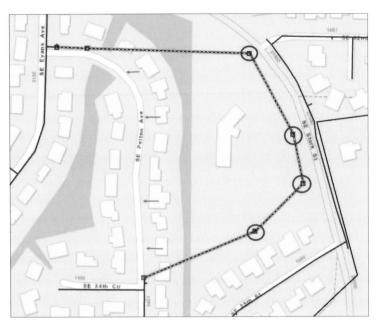

4. Finish the modification by pressing F2 on your keyboard.

5. Clear the selection.

Finally, you will remove a well feature that has run dry. To delete the feature, you will select the well and click Delete.

6. Zoom out slightly to see the well located in Sweetbriar Park, just southwest of the waterline you redrew.

7. Select the well feature on the map, and on the Edit tab of the ribbon, click Delete. Click Yes to confirm the deletion.

Saving your edits will permanently delete the well from the database.

ON YOUR OWN

Delete the waterline that led to the well.

ON YOUR OWN

Set a visibility scale for the Valves layer so that the valves are visible only at scales larger than 1:10,000.

REMIND ME HOW
- In the Contents pane, right-click Valves and click Properties.
- On the General tab, under Out Beyond (minimum scale), click 1:10,000.

8. Save your edits and the project.

You are finished with this chapter. If you are not continuing to chapter 5, you can exit ArcGIS Pro.

Summary

Although a great deal of spatial data is circulating in the world, countless scenarios exist in which features must be deleted, modified, or created from scratch. You now have a foundational skill set for such tasks. In this chapter, we did not use many additional edit tools—for instance, we did not create any curved geometry—but now you should feel comfortable enough in the editing environment to experiment on your own. Remember, all edits are temporary until saved, and you can use the Undo button to fix mistakes, so do not be afraid to develop your skills through trial and error.

Glossary terms

shapefile

geodatabase

feature class

feature dataset

coordinate system

geographic coordinate system

spheroid

projected coordinate system

map projection

on-the-fly projection

metadata

attribute domain

alias

snapping

vertex

endpoints

edge

intersection

feature template

## CHAPTER 5:

Facilitating workflows

Exercise objectives

5a: Manage a repeatable workflow using tasks
- Set up a project.
- Proceed through preconfigured tasks.
- Author a task.

5b: Create a geoprocessing model
- Define the data.
- Add operations to ModelBuilder.
- Fill out the tool parameters.
- Run the model.
- Convert a model to a geoprocessing tool.

5c: Run a Python command and script tool
- Define the data.
- Run a command using Python.
- Use a custom script tool.
- Package the project.

You are now gaining a solid understanding of how you can use GIS to manage spatial data and solve spatial problems. In this chapter, you will learn three ways for storing and automating multiple-operation workflows, an important capacity enabled through ArcGIS software. Workflow definition, storage, and automation facilitate and standardize geospatial solutions. These capabilities are especially useful for organizations that anticipate performing a workflow many times. And you can change these workflows, including the input data and other parameters, as needed. Think of the workflow processes that you will learn about in this chapter as flexible, smart, "living" documentation.

First, you will learn about automating tasks. This feature allows you to walk the user through a series of predefined steps that can incorporate (or call) commands and geoprocessing tools, including model and script tools. A task item, composed of a series of steps, might capture an entire workflow or one piece of a more complex solution. Tasks promote best practices, improving efficiency and quality. They can serve as a workflow tutorial—a documented series of instructions that call forth the correct commands or tools. And the tasks can be shared as part of a project, allowing users to share their knowledge.

Second, this chapter will introduce you to ModelBuilder™, a geoprocessing environment that allows you to easily link one tool to another and run a set of operations one after another with the click of a button. ModelBuilder provides a helpful visual diagram of geoprocessing workflows and includes advanced capabilities such as looping and if-then scenarios (although we will not get that far in this book).

Finally, you will begin to learn how to automate your work with Python, the scripting language that is compatible with ArcGIS software. You will write a command and use a custom script tool (created with Python) that performs multiple operations.

Tasks, models, and Python scripts are often used together. For example, a task step might entail running a geoprocessing tool, which might be a custom script built using Python or model tool built using ModelBuilder. It is helpful to understand all three environments, at least minimally, so that you can determine how to incorporate them into your workflows. You can use the three environments to reproduce your workflows and share them with your organization and the ArcGIS community.

Scenario: You are employed by a nonprofit agency whose primary objective is to raise awareness of and develop solutions for international human rights violations. Taking advantage of the public information offered by the Armed Conflict Location and Event Data (ACLED) project, you are creating maps that convey the impact of political violence in developing countries. You will present your maps at an international human rights conference for eventual incorporation into an awareness campaign.

The ACLED dataset you are using provides information on political violence in developing countries. Attributes include dates of conflict events, key actors involved (such as military forces, protesters, rebel groups, civilians, and others), and event classification (such as riots and protests, battles, nonviolent activities by conflict actors, violence against civilians, and more). Your initial project focuses on the African continent. Because you intend to make many similar maps for various African countries, you are exploring automation capabilities in ArcGIS Pro.

To get the data in its current state, the tabular data was downloaded from www.acleddata.com to an Excel file. Rows of data recorded prior to the year 2005 were deleted; the spreadsheet was saved as a .csv file and converted to a point feature class using the Table to Point tool in ArcGIS Pro.

GIS IN THE WORLD: MAPPING THE WALKABILITY OF CANADIAN NEIGHBORHOODS

A community designed to promote more walking and bicycling and less driving also benefits citizens' health and the environment. The Halton Region's Planning Services Division, along with the region's Health Department, spearheaded a GIS project to measure, quantify, model, and map the walkability of a large region of Ontario, Canada. They used ArcGIS and the ArcGIS Network Analyst™ extension for data management and analysis. They also used ModelBuilder to automate the analysis so that it could be easily modified and repeated for other regions.

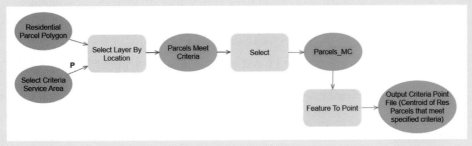

Figure 5.1. The project uses several models. This is an example of a model for selecting residential parcels within walking distance of a specified diverse-use (DU) development.

Read about the project in *ArcUser*, "Modeling Walkability":
www.esri.com/news/arcuser/0112/modeling-walkability.html.

DATA

In GTKAGPro\Conflict\Conflict.gdb:

- ACLED_2005_2018–a point feature class that displays locations of armed conflicts throughout Africa from 2005 to 2018 *(Source: ACLED)*

Exercise 5a: Manage a repeatable workflow using tasks

Estimated time to complete: 30 minutes

In this exercise, you will begin using the Tasks pane, an ArcGIS Pro feature that allows you to follow or define a workflow for yourself or other users. A *task item* is a predefined series of steps, grouped into tasks, that walk you through a GIS workflow. Each *task* contains one or more related steps. Tasks are helpful to standardize business operations and promote best practices for a repeatable workflow. As you will see, some of the steps can be automated or semiautomated, saving time and reducing the possibility of user error.

Exercise workflow

- Add the ACLED data for the African continent.
- Examine the dataset's attributes.
- Follow preconfigured tasks to do the following:
  - Limit the layer's definition so that it shows only conflicts in a single country during specific years.
  - Rename and symbolize the defined layer.
  - Select only those conflicts classified as "violence against civilians."
- Create a new task to summarize the number of fatalities within the selected set of features.

Set up a project

1. In ArcGIS Pro, open Conflict.aprx from the GTKAGPro\Conflict folder.

    So far, the map contains only a topographic basemap layer, centered on the African continent.

2. Examine the Catalog pane by expanding the Tasks folder, Conflict folder, Conflict.gdb, and Conflict.tbx, as shown in the figure.

    The project includes two layer files, a geodatabase (Conflict.gdb) that contains a point feature class, a toolbox that includes a custom script tool, and a task item named Create Conflict Maps.

- ▷ ⬚ Maps
- ▷ ⬚ Toolboxes
- ▷ ⬚ Databases
- ▷ ⬚ Styles
- ◢ ⬚ Tasks
  - ⬚ Create conflict maps
- ◢ ⬚ Folders
  - ◢ ⬚ Conflict
    - ◢ ⬚ Conflict.gdb
      - ⬚ ACLED_2005_2018
    - ◢ ⬚ Conflict.tbx
      - ⬚ Select and Summarize
    - ⬚ ACLED_2010_2018_Rwanda.lyrx
    - ⬚ ACLED_2010_2018_SouthSudan.lyrx
- ▷ ⬚ Locators

First, you will add your project data to the map.

3. Drag ACLED_2005_2018 from the Catalog pane to the map. (Your color symbology may not match the image in the book.)

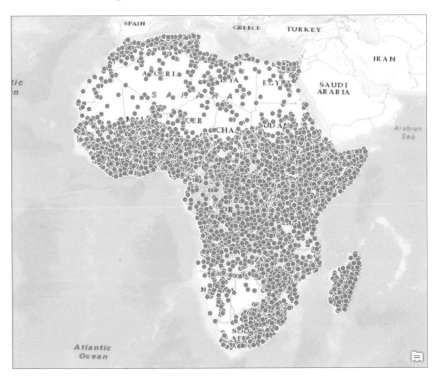

4. Open the layer's attribute table and examine the field values.

YEAR	TIME_PRECISION	EVENT_TYPE	ACTOR1	ASSOC_ACTOR_1	INTER1
2005	1	Battle-No change of territory	GSPC: Salafist Group fo	<Null>	2
2005	1	Strategic development	GSPC: Salafist Group fo	<Null>	2
2005	1	Battle-No change of territory	GSPC: Salafist Group fo	<Null>	2
2005	1	Riots/Protests	Rioters (Algeria)	<Null>	5
2005	1	Riots/Protests	Protesters (Algeria)	<Null>	6

Can you name the types of conflict events that are recorded in this dataset?

5. Close the attribute table.

Next, you will follow a task to perform a repeatable workflow.

Proceed through preconfigured tasks

1. On the View tab, in the Windows group, click the Tasks button to show the Tasks pane.

When the Tasks pane opens, you will see the Create Conflict Maps task item. The task item includes three separate tasks: Create Subset of Features, Symbolize Layer, and Make Selection.

2. Point to the first task to reveal the blue arrow. Click the arrow to open the task.

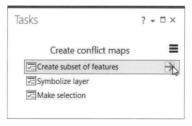

The first task contains three steps. The Progress indicator at the bottom says that you are on step 1 of 3. Familiarize yourself with the layout of the Tasks pane.

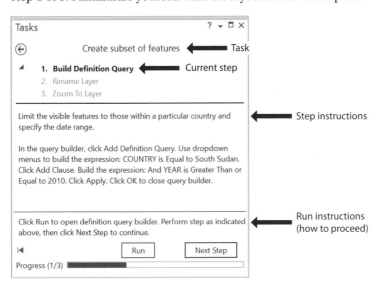

The first step facilitates the creation of a *definition query*, which limits the display of features to features that meet user-defined criteria.

3. Read the step and instructions for step 1 of the task.

   If necessary, expand the Tasks pane to see all the instructions.

4. Click Run.

   This step opens the query builder, which is on the Definition Query tab of Layer Properties. The rest of the step relies on manual input, with the instructions documented in the task step.

5. Carefully follow the step instructions in the Tasks pane to build the *query expression*. After you build the expression, be sure to click Apply and then OK.

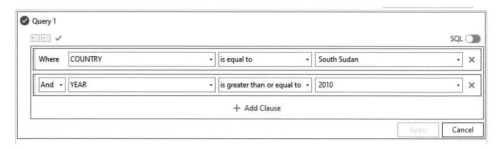

After the definition query is run, the map displays fewer features.

6. After you close the query builder, click Next Step to proceed to step 2: Rename Layer.

7. Click Run to open the layer properties, follow the step instructions to rename the layer, and then click Next Step.

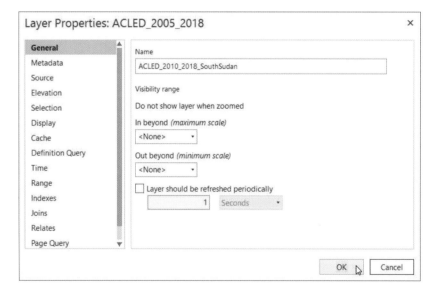

Step 3 of the task was configured to run automatically. Your map zooms to the extent of the feature points.

8. Click Finish to complete the task.

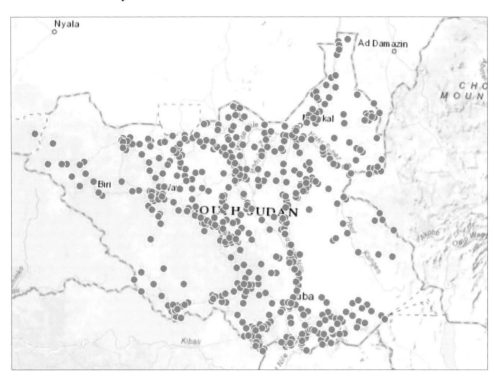

9. Open the next task, Symbolize Layer.

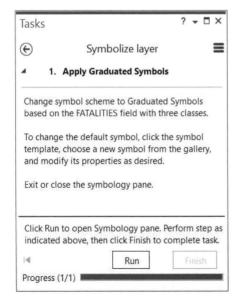

This task has only one step (you can tell because the Progress indicator says 1/1). The task author can give a task as few or as many steps as required.

10. Now that you are familiar with the way the Tasks pane works, complete the Symbolize Layer task on your own.

When you click Run, the Symbology pane appears.

11. Follow the task step instructions to configure symbols.

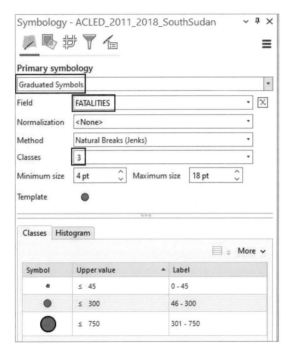

The next task calls a geoprocessing tool.

12. When you have finished the Symbolize Layer task, open the Make Selection task.

This task also has one step. The Select Layer by Attribute tool is embedded within the Tasks pane, with default parameters preselected. The selected features will be features with an EVENT_TYPE field value of Violence against Civilians (point to the expression to see it in its entirety).

13. Make sure that Selection Type is set to New Selection and that the expression is correct.

You can change geoprocessing tool parameters embedded in a task.

14. Click Run. When the process is complete, click Finish, but keep the Tasks pane open.

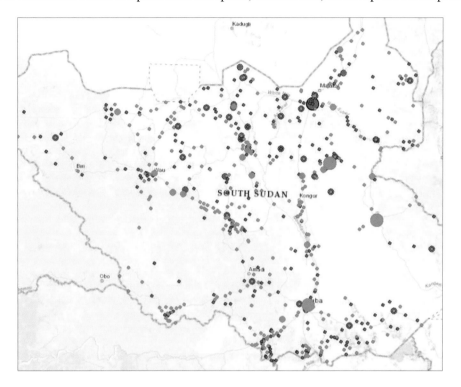

The map shows conflicts in South Sudan between 2010 and 2018, symbolized with graduated symbols, so that the increasing size corresponds to a greater number of fatalities. Conflicts that have been classified as violence against civilians are selected in cyan.

Author a task

Next, you will add to an existing task item.

1. In the Tasks pane, click Create Conflict Maps, and click the Options menu > Edit in Designer.

This step opens the Task Designer. You will be working with both windows simultaneously, so you will want to read the upcoming instructions carefully; they will be explicit about whether you should be in the Tasks pane or the Task Designer.

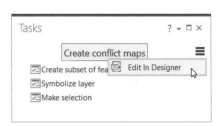

When the Task Designer is open, the Tasks pane reveals new buttons and tabs.

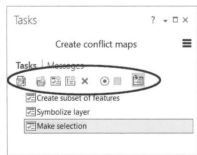

2. In the Tasks pane, click the New Task button.

A new task is now visible in the Tasks pane, and the Task Designer is updated to reflect the new task. Clicking any task in the Tasks pane will bring up the same task in the Task Designer.

When you have the Task Designer open, you can reorder tasks by dragging them in the Tasks pane.

3. In the Task Designer, name the new task **Generate summary statistics**. For description, type **Summarize the fatalities for the selected set of features.**

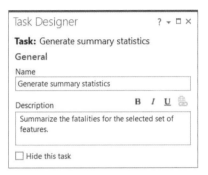

This task item will have one step. To populate the step, you will record your actions.

4. In the Tasks pane, click the Record Commands button.

Your ArcGIS Pro interactions are now being recorded and will be translated into an automated step when you stop recording.

If you want to enter a step manually, without using the Record Commands option, click the New Step button.

5. On the ribbon, click the Analysis tab. In the Tools gallery, find and click the Summary Statistics tool. (You can also open the Geoprocessing pane and search by tool name.)

6. Do not fill out the tool. Close the Geoprocessing pane (you can also keep it open if you prefer).

7. In the Tasks pane, click the Stop Recording button.

8. Back in the Task Designer, for Run/Proceed Instructions, type **Click Run to run geo-processing tool. Click Finish to complete task.** Maintain the default step behavior (Manual).

9. In the Task Designer, click the Actions tab. Point to Summary Statistics, and click the Edit tool. Set the parameters as follows, and then click Done.
    - Input Table: ACLED_2010_2018_SouthSudan
    - Output Table: Conflict.gdb**ACLED_SouthSudan_Statistics**
    - Field: FATALITIES
    - Statistic Type: Sum

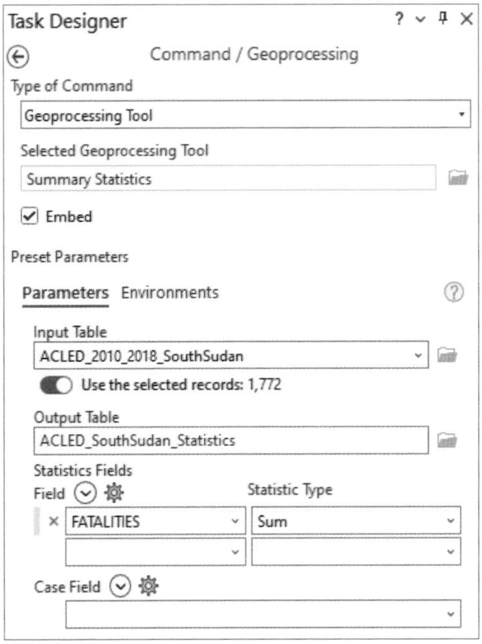

10. Close the Task Designer.

11. Run the Generate Summary Statistics task, and then click Finish to complete the task item.

12. Open the new statistics table that has been added to the Contents pane.

According to ACLED statistics, how many fatalities resulted from violent conflicts against South Sudanese civilians between 2010 and 2018?

To see the new table in the Catalog pane, right-click Conflict.gdb and click Refresh.

ON YOUR OWN
Repeat the entire task for another African country of your choice.

13. Close the table and the Tasks pane, clear the selection, and save the project.

> **ON YOUR OWN**
>
> In the Catalog pane, under Tasks, right-click Create Conflict Maps, and click Export To File > Save As. Name the task file as you like, and save it to your Conflict folder. You can now share the file with a colleague by attaching it to an email or putting it in a network. Saving the task file is a good way to share a workflow when you do not want to share an entire ArcGIS Pro project.

Tasks provide the benefit of a built-in, how-to tutorial without needing separate documents to explain workflows. Whether they are commonly performed tasks or more obscure workflows that are performed only occasionally, storing the sequence of steps in a task item is a good way to ensure compliance and consistency.

For more information on tasks, in ArcGIS Pro Help, go to Workflows > Tasks.

You can continue to the next exercise, or exit ArcGIS Pro and come back to it later.

Exercise 5b: Create a geoprocessing model

Estimated time to complete: 25 minutes

ModelBuilder, packaged within ArcGIS Pro, is a design environment for creating spatial analysis workflow diagrams. The diagrams you make are called models, and they have many advantages. A *model* allows you to string multiple geoprocessing tools together and run them automatically with the click of a button. The output of one process becomes the input for the next process, and so on.

In this exercise, you will perform a workflow like the one you did in exercise 5a; however, instead of using tasks, you will use ModelBuilder to build, save, and automate the workflow.

Exercise workflow

- Create a new ACLED layer with a limited feature definition.
- Build a model to do the following:
    - Apply the symbology from one layer to another.
    - Select only those conflicts classified as violence against civilians.
    - Summarize the number of fatalities within the selected set of features.

Define the data

Before you jump into ModelBuilder, you will configure a new feature layer to serve as the model's first input dataset.

1. In ArcGIS Pro, continue working with the Conflict project that you modified in exercise 5a (GTKAGPro\Conflict\Conflict.aprx).

   The project should have a layer named ACLED_2010_2018_SouthSudan, which displays conflicts that occurred in South Sudan between 2010 and 2018. Conflict locations are symbolized using graduated symbols so that larger symbols show greater fatalities. The Contents pane also has a statistics table.

   If you did not successfully complete exercise 5a, add the layer file ACLED_2010_2018_SouthSudan.lyrx from the project folder.

   Recall that layer files are not actual data; they point to a source dataset (in this case, the ACLED_2005_2018 geodatabase feature class), but they provide layer properties, such as definition queries, selections, or symbology schemes. You do not need the statistics table.

2. From Conflict.gdb, add the ACLED_2005_2018 feature class to your map, and zoom to the layer.

   This complete layer will nearly cover the Sudan layer.

3. Open the properties of the new layer, and click the Definition Query tab.

4. Build two definition query clauses, and click Apply:
- Where COUNTRY is Equal to Rwanda
- And YEAR is Greater Than or Equal to 2010

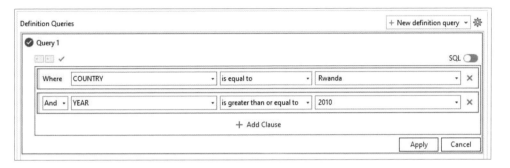

REMIND ME HOW
- Click New Definition Query.
- In the first drop-down box, scroll down and click COUNTRY. Maintain the default selection (is Equal to) for the second drop-down box. For the third drop-down box, click Rwanda.
- Click Add Clause to finalize the first query.
- Configure the next four drop-down boxes to build the next query statement: And YEAR is Greater Than or Equal to 2010.

5. Click the General tab. Change the layer name to **ACLED_2010_2018_Rwanda,** and then click OK.

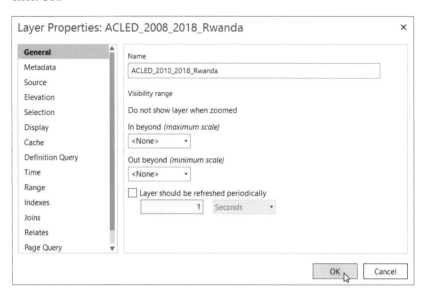

6. Zoom to the extent of the new layer.

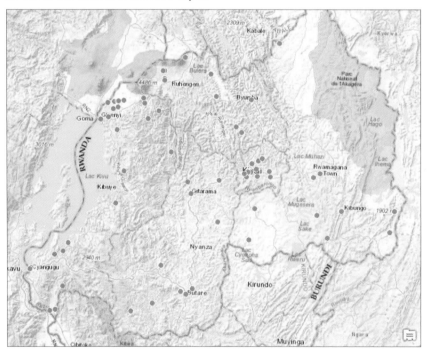

7. Symbolize ACLED_2010_2018_Rwanda using graduated symbols based on the FATALITIES field. Use the default classification method but change the number of classes to **3**, and modify the symbol template to your liking.

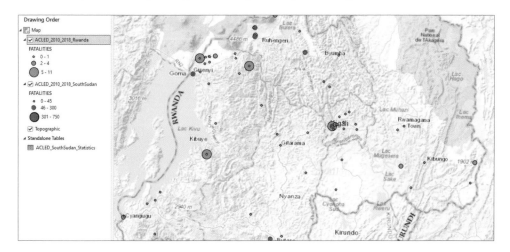

REMIND ME HOW

- To open the Symbology pane, with the new layer selected in the Contents pane, click the Symbology button on the Feature Layer tab.
- In the Symbology pane, choose Graduated Symbols from the Primary Symbology drop-down list.
- Change the Field option to FATALITIES.
- Maintain the default classification options, and change the Classes to **3**.
- To change the appearance of the points, click the symbol template.
- Choose a new symbol from the Gallery.
- Modify the properties as desired.

Add operations to ModelBuilder

Next, you will create a new geoprocessing model that will reside in your project's toolbox.

1. On the Analysis tab, in the Geoprocessing group, click the ModelBuilder button.

    This step opens the ModelBuilder view. By default, it is docked in the map view area, on a new tab. You can undock the window if you prefer.

    A ModelBuilder tab is now available on the ribbon.

    Alternatively, you can right-click a toolbox in the Catalog pane and click New > Model.

2. In the Catalog pane, expand Toolboxes > Conflict.

    Notice that the Conflict project toolbox contains a new empty model. (This model was automatically created when you opened the ModelBuilder window.) The toolbox also contains a script tool, which you will use in exercise 5c.

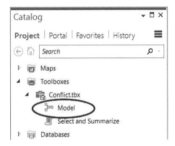

3. In the Contents pane, drag the ACLED_2010_2018_Rwanda layer to the empty ModelBuilder window.

    A blue oval with the name of the layer appears in the Model view. You will use this layer as the input for a geoprocessing tool.

4. On the ModelBuilder tab, in the Insert group, click the Tools button to open the Geoprocessing pane, if necessary.

5. In the Geoprocessing pane search box, type **Select**, and press Enter. From the search results, drag the Select Layer by Attribute tool to the Model window.

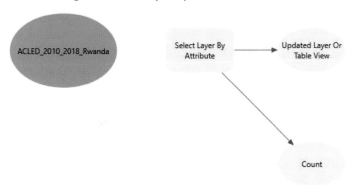

6. In the same way, search for **Summary**, and drag the Summary Statistics tool to the ModelBuilder window.

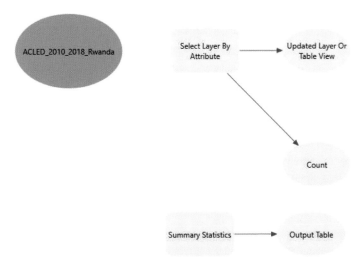

You have two empty tools in your model (with elements in light gray). Next, you will define the tool parameters.

Fill out the tool parameters

1. Click ACLED_2010_2018_Rwanda, and drag an arrow onto the Select Layer by Attribute tool element, release, and click Input Rows to add the layer as input.

2. Double-click the Select Layer by Attribute tool.

   The Input Rows parameter is already set to ACLED_2010_2018_Rwanda because you added an arrow.

3. In the tool pane, keep the default selection type (New Selection). Under Expression, build the following expression: Where EVENT_TYPE is Equal to Violence against civilians.

4. Click OK.

   You can increase the size of the elements in your model window by adjusting the size bar at the bottom.

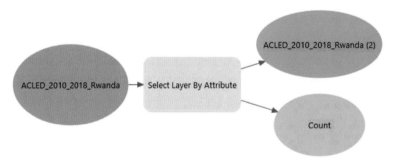

   The tool in the model shows an input layer with a color, which indicates that it is ready to run.

   To see the entire input and output layers in the model, click an element and resize it by dragging the selection anchors. If the ModelBuilder window is floating, resize it as necessary.

5. Draw an arrow from the green output, ACLED_2010_2018_Rwanda (2), to the Summary Statistics tool element, release, and click Input Table.

6. In the model, double-click the Summary Statistics tool.

For Input Table, the parameter is now set to ACLED_2010_2018_Rwanda. You are choosing intermediate data; that is, the output of the first operation becomes the input of the second operation. Do not worry if the text box still shows Rwanda.

7. For Output Table, keep the default location, but replace the default name with **ACLED_ Rwanda_Statistics**. For Field, click FATALITIES. For Statistic Type, click Sum. Leave the Case field parameter blank. Click OK.

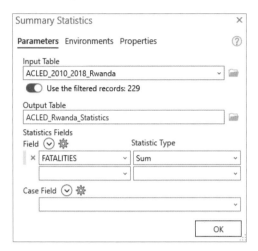

Using ModelBuilder, you have connected this operation to the output of the selection operation. The statistics will be calculated only for selected features.

8. On the ModelBuilder ribbon, in the View group, click the Auto Layout button. Resize the elements or the ModelBuilder window as needed. If you want, click the Fit to Window button or adjust the size bar at the bottom.

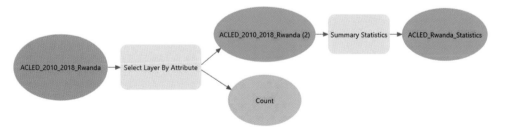

9. On the ModelBuilder tab, in the Model group, click Save to save the model.

    Alternatively, click Save As, and give the model a unique name, such as
    ACLED_select_stats.

You can run tools in a model one by one (right-click the tool and click Run), or run connected tools automatically, which is what you will do next.

Run the model

1. On the ModelBuilder tab, in the Run group, click Run.

   A progress window appears to indicate the model's status. You may also notice that each tool in the model turns red as it is processing.

2. When the model has completed all the processes, close the progress window.

3. Switch to the Map view to see the map.
   In the Rwanda layer, conflicts classified as violence against civilians are selected in cyan.

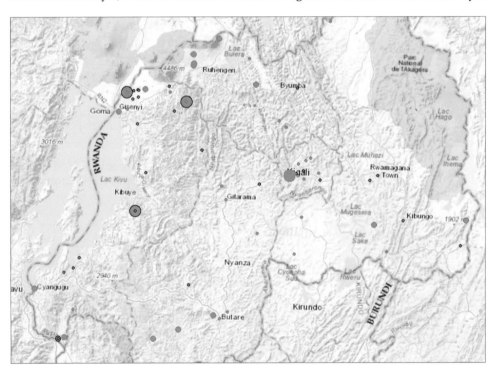

4. In the Catalog pane, expand Conflict.gdb, and refresh the view (right-click and click Refresh).

The geodatabase now contains a new table, ACLED_Rwanda_Statistics.

5. Drag the table to the map, and open it to review the data.

How many fatalities resulted from conflicts classified as violence against civilians in Rwanda from 2010 to 2018?

This model was a simple one. Models can be as simple or as complex as needed. The value of a model is that you can easily modify the input dataset and any tool parameters as necessary; you can save the model with a new name if desired and run the model on a new dataset. You can tweak, refine, and add to the model whenever you want, without having to start from scratch. You can also set model parameters so that the model can be run from a geoprocessing tool dialog box, which you will do next.

6. Close the table when you have finished.

Convert a model to a geoprocessing tool

You can turn a model into a stand-alone geoprocessing tool that you can share and then repeat the process, including changing the input, output, and variables.

1. In ModelBuilder, right-click the input element (the blue oval), and click Parameter. The letter *P* is added near the element.

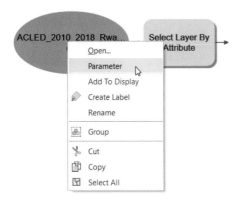

2. Right-click the Select Layer by Attribute tool element (the first yellow rectangle), and click Create Variable > From Parameter > Expression.

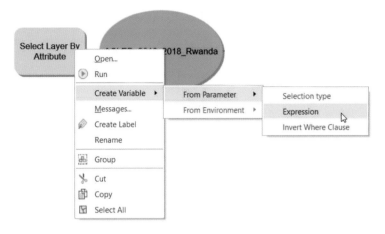

3. Right-click the Summary Statistics tool element, and click Create Variable > From Parameter > Statistics Fields.

4. Set the final output and both variables as parameters.

5. Move the variable elements as you want, or click the Auto Layout button.

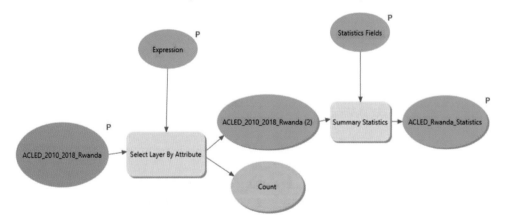

6. Save the model, and exit ModelBuilder.

7. In the Catalog pane, find the model tool in the Conflict toolbox, and double-click it to open it.

Your model looks like a regular geoprocessing tool.

8. Change the output variable to **ACLED_Rwanda_Statistics2,** edit the Event type to Riots/Protests, and then run the tool.

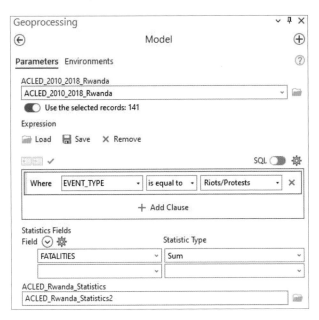

The new statistics table is automatically added to the map.

How many events classified as riots/protests occurred in Rwanda between 2010 and 2018? How many fatalities resulted from these events?

You can also repeat the first seven steps of this exercise ("Define the data"), choosing a different country to display, and then choose that new layer as your model input, varying the selection and statistics field as desired.

9. Clear the selection, close the model and any open tables, and save the project. You will continue using the model in exercise 5c.

Exercise 5c: Run a Python command and script tool

Estimated time to complete: 25 minutes

You now have experience running geoprocessing functions in two ways: using stand-alone tools or combining them in the ModelBuilder window.

You can also run geoprocessing tools, along with other functions, using Python. *Python* is a programming language that, in the GIS context, is used to *script* geoprocessing workflows and build custom geoprocessing tools.

You can type Python commands from the Python window (*command line*) in ArcGIS Pro. ArcPy is the name of the Python site package that allows you to run ArcGIS commands from Python.

A full synopsis of Python and coding techniques is beyond the scope of this book. Think of this exercise as a first look at the Python window in ArcGIS Pro, something you may want to explore in depth later. Even if you are well-schooled in coding, this exercise will be helpful to see how you can incorporate Python into ArcGIS Pro.

You will follow a now familiar workflow, similar to the workflow in exercises 5a and 5b. But this time, you will use Python coding and a script tool to run the processes.

Exercise workflow

- Create a new ACLED layer using a limited feature definition and meaningful symbology.
- Use a geoprocessing tool and then Python code to do the following:
  - Create a layer file (.lyrx) from the new map layer.
- Use a custom script tool to do the following:
  - Select only those conflicts classified as violence against civilians.
  - Summarize the number of fatalities within the selected set of features.

Define the data

Before jumping into what might be your first experience in writing code in ArcGIS software, you will set up a new map layer.

1. Continue working in Conflict.aprx.

It should contain two ACLED layers (ACLED_2010_2018_SouthSudan and ACLED_2010_2018_Rwanda), both symbolized using graduated point symbols based on the FATALITIES field.

If you did not successfully complete exercises 5a and 5b, add the layer files ACLED_2010_2018_SouthSudan.lyrx and ACLED_2010_2018_Rwanda.lyrx from the GTKAGPro\Conflict folder.

2. Add another copy of the ACLED_2005_2018 feature class to the map. Limit the layer's definition to the country of Nigeria, displaying conflicts between 2010 and 2018.

3. Rename the layer **ACLED_2010_2018_Nigeria**, and zoom to the defined layer.

4. Symbolize ACLED_2010_2018_Nigeria using graduated symbols based on the FATALITIES field. Use the default classification method, change the number of classes to **3**, and change the symbols to whatever you prefer.

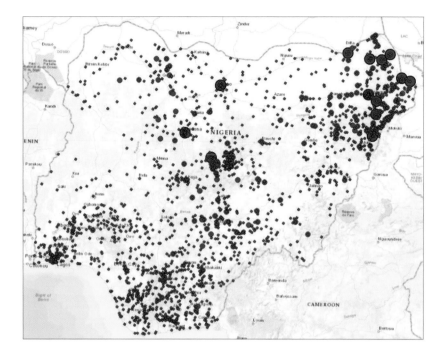

Run a command using Python

Next, you will create a layer file to store the definition query, name, and symbology scheme of this layer. Then it can be used in other maps or projects without having to repeat the preceding steps. Layer files also save label characteristics and custom layer pop-up windows, if applicable. You will first run this command using the Save to Layer File geoprocessing tool, again using copied Python code, and a third time by entering the code yourself.

1. On the Analysis tab, click the Tools button to open the Geoprocessing pane.

2. Type the search word **save**, and open the Save to Layer File tool.

The tool parameters require an input layer, an output file name, and an optional setting to save the file using a relative path. (Using this option means that if you move the layer file to another location, its path to the source data is updated accordingly.)

3. Set the following tool parameters, and click Run:
- Input Layer: ACLED_2010_2018_Nigeria.
- Output Layer: GTKAGPro\Conflict**ACLED_2010_2018_Nigeria.lyrx**.
- Check the check box to store relative paths.

4. When the tool is finished running, save the project.

5. Right-click the status bar at the bottom of the Geoprocessing pane, and click Copy Python Command.

If you already closed the Geoprocessing pane, you can alternatively copy the Python command from the geoprocessing history on the Analysis tab > Geoprocessing group.

6. On the View menu, click the Python Window button.

This step opens the Python window. Like other windows and panes, it can be docked in one of several locations or be free-floating.

7. At the bottom of the Python window, right-click in the code entry box, and click Paste.

All geoprocessing tools are accessible from ArcPy. If you have a workflow that uses 10 geoprocessing tools, you can write 10 lines of code, with one line calling each tool.

8. Click at the end of the copied code to see a code parameter description pop-up message.

Just like the geoprocessing tool, the Python code parameters expect you to enter an input layer and output layer. Optionally, you can enter a parameter to set the relative path option and choose which version of the data to use when multiple versions exist. Optional parameters are marked with curly brackets {like this} in the description pop-up message.

If you run the code as is, the operation will fail because there is already a layer file in the Conflict folder named ACLED_2010__2018_Nigeria.lyrx—the one you just created from the Geoprocessing pane.

9. In the Python window, change the output file name to ACLED_2010_2018_Nigeria**2**. lyrx.

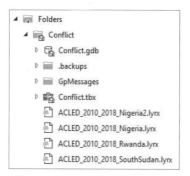

```
arcpy.management.SaveToLayerFile(
    in_layer="ACLED_2010_2018_Nigeria",
    out_layer=r"C:\GTKAGPro\Conflict\ACLED_2010_2018_Nigeria2.lyrx",
    is_relative_path="RELATIVE",
    version="CURRENT"
)
```

10. Again, click at the end of the code, and then press Enter to run the command. When processing is complete, a second Nigeria layer file is added to the Conflict folder.

```
▲ 📷 Folders
    ▲ 📷 Conflict
        ▷ 📷 Conflict.gdb
        ▷ 📁 .backups
        ▷ 📁 GpMessages
        ▷ 📷 Conflict.tbx
            📄 ACLED_2010_2018_Nigeria2.lyrx
            📄 ACLED_2010_2018_Nigeria.lyrx
            📄 ACLED_2010_2018_Rwanda.lyrx
            📄 ACLED_2010_2018_SouthSudan.lyrx
```

If you have time, you will now run the Save to Layer File tool a third time by manually entering the code.

11. To clear the history in the Python window, right-click the completed code, and click Clear Transcript.

The following steps are optional. At least read through them so you understand that there are additional steps required if you are working outside ArcGIS or your desired geoprocessing workspace.

12. In the code entry portion of the Python window, type the following: **import arcpy**.

13. Press Enter to add the line of code to the upper portion of the Python window.

This line is a matter of protocol. In ArcGIS applications, it is not required. But if you want to run Python commands outside ArcGIS, it is required, so adding the line is good practice.

Next, you will set the environment to ensure that the operation's output will be stored where you want it. Again, this step is optional, because you are already in the workspace you are calling, but if you are working in a different project or environment, it is required.

14. Type the following code, and then press Enter: **arcpy.env.workspace = 'C:/GTKAGPro/ Conflict'**

Single quotation marks and double quotation marks are treated the same; for example, you can enter 'name' or "name" as code. Also, paths are not case sensitive. You can also type it this way: `arcpy.env.workspace = "c:/gtkagpro/conflict"`

```
import arcpy
arcpy.env.workspace = 'C:/GTKAGPro/Conflict'
|
```

15. Begin to type the following line: **arcpy.SaveToLayerFile**.

Notice that the Python window offers the correct entry when you begin typing.

16. Click the auto-complete suggestion to call the tool.

The highlighted pair of parentheses after the tool and toolbox name is where you add your tool parameters.

17. With the pointer between the parentheses, notice that you are offered a list of map layers as possible inputs. Click ACLED_2010_2018_Nigeria.

```
arcpy.management.SaveToLayerFile('ACLED_2010_2018_Nigeria')
```

You can also drag inputs from the Contents pane.

18. After the input dataset (following the quotation mark but inside the closing parenthesis), type a comma and then a single quotation mark. Notice that the closing quotation mark is provided for you. Type the output layer file name: **ACLED_2010_2018_ Nigeria3.lyrx**

```
arcpy.management.SaveToLayerFile('ACLED_2010_2018_Nigeria','ACLED_2010_2018_Nigeria3.lyrx')
```

19. After the output quotation mark, type another comma, and then from the options box click either 'RELATIVE' or 'ABSOLUTE'.

20. Type another comma, and then choose the only available option ('CURRENT').

```
arcpy.management.SaveToLayerFile('ACLED_2010_2018_Nigeria','ACLED_2010_2018_Nigeria3.lyrx', 'RELATIVE', 'CURRENT')
```

The command is ready to run.

21. Click after the closing parenthesis, and press Enter.

22. When the process is complete, find the new layer file in the project folder.

23. Close the Python window.

You have learned how to use Python by copying existing code and typing it manually. You can delete the redundant layer files if you want.

Use a custom script tool

Script tools look like any ArcGIS geoprocessing tool, except that they are based on a Python script. You can use Python to create a custom geoprocessing tool. Similar to creating a geoprocessing tool based on a model, the tool author specifies the tool parameters that the end user can modify.

1. In the Conflict toolbox, notice the Select and Summarize custom script tool. To view the code behind the tool, right-click the tool, and click Edit.

```
# Select and Summarize
# -----------------------------------------------------------------------
# Created on: 2020-06-17
#
# Description: The user can select an input file to symbolize and an output file to store the summary results
#
# -----------------------------------------------------------------------

# Import arcpy module
import arcpy
from arcpy import env

# Set the environment for output file location
env.workspace = ("c:\GTKAGPro\Conflict\Conflict.gdb")

# Prompt the user for the name of the input layer
ACLED_2005_2018_InputFile = arcpy.GetParameterAsText(0)

# Prompt the user for the name of the output file
ACLED_2010_2018_Output_Stats = arcpy.GetParameterAsText(1)

# Process: Select Layer By Attribute
arcpy.SelectLayerByAttribute_management(ACLED_2005_2018_InputFile, "NEW_SELECTION", "EVENT_TYPE = 'Violence against civilians'",

# Process: Summary Statistics
arcpy.Statistics_analysis(ACLED_2005_2018_InputFile, ACLED_2010_2018_Output_Stats, "FATALITIES SUM", "")
```

Anything preceded by a hashtag (#) is a comment and is ignored by Python; it is for the user's benefit.

The code includes the tool name, creation date, and description, followed by the processing instructions: the call to import ArcPy, environment settings, geoprocessing tool parameters, and the geoprocessing tools run by the script.

Which geoprocessing tools are combined in this script?

If your data is not stored in C:\GTKAGPro, you should manually edit the "env.workspace = " line, and then click File > Save.

2. Close the Select and Summarize script window.

3. In the Catalog pane, right-click the Select and Summarize script tool, and click Open.

You saw the back end of the tool a moment ago—the code behind it. This view is the front end, and it looks like any other geoprocessing tool. But instead of performing one process, this tool performs two, and it can be more than two. Script tools are especially good for automated workflows and batch processing.

4. For Select Input Layer, select ACLED_2010_2018_Nigeria.

    The selection will be made in this layer. Recall that the second process, after the attribute selection, is to create a table of summary statistics based on the total fatalities within the selected set (violence against civilians).

5. For the output file name, type **ACLED_Nigeria_Statistics**.

    The tool is already programmed to store the table in the Conflict.gdb workspace.

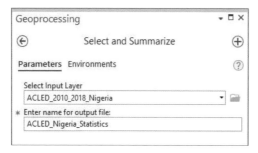

6. Run the tool.

    Notice that the Nigeria conflicts that are characterized as violence against civilians have been selected, as indicated by the script tool code. The tool also created a table that summarizes the Fatalities value of the selected set.

7. Find the new table in the Catalog pane. (Refresh the geodatabase if necessary.) Drag the table to the map, and open it.

*How many fatalities resulted from conflicts classified
as violence against civilians in Nigeria from 2010 to 2018?*

This introduction to Python is only cursory. You began and did some programming, which is great, even if you did not fully understand the underlying concepts of Python. Many resources are available that can give you more thorough lessons in using Python with ArcGIS Pro. For starters, explore the ArcGIS Pro ArcPy Reference at http://pro.arcgis.com.

If you choose to share your saved project, you are also sharing the task, model, and script tool that you created in this chapter.

8. Close the table, clear the selection, and save the project.

Package the project

To share your project, including the task, model, and script tool you created, you must create a project package.

1. Click the Share tab, and click the Project button.

2. In the Package Project pane, maintain the option to Upload Package to Online Account.

3. Maintain the default name. Under Summary, enter a sentence or two that explains the project.

4. Under Tags, enter some words that you think capture the key points about the project (such as **conflict**, **Africa**, and **ACLED**).

5. Save the project to your root folder, and choose to share it with your organization or selected groups. Click Package.

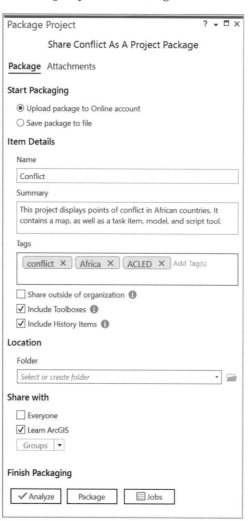

ON YOUR OWN

Sign in to ArcGIS Online (www.arcgis.com). Under Content, find the project.

You are finished with this chapter. If you are not continuing to chapter 6, you can exit ArcGIS Pro.

Summary

You now have some experience using ArcGIS Pro tasks, ModelBuilder models, and Python scripting to automate GIS workflows. As you continue using GIS in your own profession, you may find that one environment is best suited for solving a problem, or you may find that you can incorporate all three environments. Now that you have taken a first look at each method, you should feel comfortable doing more investigation and learning on your own.

Glossary terms

task item
task
definition query
query expression
ModelBuilder
model
Python
script
command line

## CHAPTER 6

Geoenabling your project

Exercise objectives

6a: Prepare project data
- Join a table.
- Symbolize using graduated colors.

6b: Geocode location data
- Create an address locator.
- Geocode addresses.
- Rematch addresses.

6c: Use geoprocessing tools to analyze vector data
- Create buffers.
- Merge and dissolve features.
- Clip features.
- Select by attribute and location.
- Create a spatial join.

Earlier in the book, you learned how to create and modify features. You can also create features from information that describes or names a location—typically an address—through a process called *geocoding*. When working on projects, you are often faced with situations in which data is not readily available. One of the advantages of geocoding is the ability to quickly create data from a list of points of interest that you have collected or researched.

You need several components to geocode addresses:

- An address table—a list of addresses for features that you want to map and that are stored as a database table or a text file
- Reference data—commonly from a street layer, on which the addresses can be located
- An *address locator*—a file that contains the reference data and various geocoding rules and settings

The ArcGIS Pro geocoding service uses address information in the attribute table of the reference data to calculate where addresses from the address table are located and creates points at those locations. As with other GIS data, the higher the quality of the reference data, the more accurately addresses can be geocoded. Ideally, street-level reference data includes attributes such as street name, street type, postal or zip code, and directional prefixes and suffixes to avoid ambiguity in address locations. And each street is divided into line segments that have starting and ending address numbers so that you can determine separate address ranges for each side of the street.

Address locators come in many styles; your reference data will determine which style can geocode your address table appropriately. In ArcGIS Pro, address locators are created in the Geoprocessing pane. For more information on geocoding, go to ArcGIS Pro Help > Tool Reference > Geoprocessing Tools > Geocoding Toolbox.

Scenario: Deciding on a retail site location is one of the most critical decisions a business must make. The success of the store depends on many factors related to location and demographics, such as access to potential and existing customers, visibility in its neighborhood, and appeal to its targeted audience, as well as wider demographics to catch potential consumers. In this chapter, you are an analyst who will prepare geographic and demographic data for use in support of a high-end bicycle retail site location and consultation service. You will use several important tools to prepare the data, which will include joining tables from other sources of data, making selections, and spatially joining data. You will also geocode a table of prospective retail sites by address and use *overlay* analysis tools to identify and visualize how potential retail site locations may serve customers in the neighborhood.

GIS IN THE WORLD: SITE SELECTION

The Corpus Christi Regional Economic Development Corporation (CCREDC) is a non-profit organization dedicated to attracting, building, and growing projects that foster economic development in and around Corpus Christi, Texas, which is situated along the Gulf of Mexico. The CCREDC created a map to show the economic development region to help potential businesses and prospects that want to relocate to the area to conduct self-guided research before getting in contact for more information. This allowed the CCREDC to respond to requests for information about the community and various properties and make detailed pitches to prospects.

Read more about how the CCREDC created a site selection solution for business prospects by incorporating data from the region in *ArcNews*, "With GIS-Based Site Selection Solution, Corpus Christi Gains a Development Edge": https://www.esri.com/about/newsroom/arcnews/with-gis-based-site-selection-solution-corpus-christi-gains-a-development-edge.

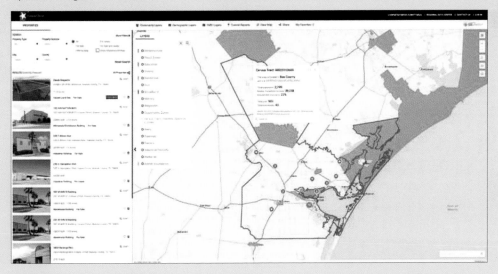

Figure 6.1. Clients can choose the layers they want to present to their users, such as federal opportunity zones, railroads, pipelines, and water and sewer lines. Courtesy of Corpus Christi Regional Economic Development Corporation.

DATA

In GTKAGPro\RetailSiteStudy\RetailSiteStudy.gdb:

- Bike Stations—a point feature class showing bike stations (*Source: City of Houston*)
- Existing Bike Lanes—a line feature class of existing bike lanes (*Source: City of Houston*)
- Proposed Bike Lanes—a line feature class of proposed bike lanes (*Source: City of Houston*)
- Commercial Land Use—a polygon feature class derived from a large land-use data-set (*Source: City of Houston*)
- Census Block Demographics—a polygon feature class of census block groups in the city of Houston (*Source: City of Houston*)
- Median household income (2010 census)—a table file that contains median house-hold income values at the census block level for the city of Houston (*Source: US Census Bureau, City of Houston*)
- Roads—a line feature class (*Source: City of Houston*)

In GTKAGPro\RetailSiteStudy:

- retail_site_prospects—a text file that contains a compiled list of commercial properties

Exercise 6a: Prepare project data

Estimated time to complete: 20 minutes

In this exercise, you will prepare census data for use later in the analysis.

Exercise workflow

- Add the median household income table to the project.
- Review table information.
- Join the median household income table to the census blocks layer.
- Symbolize census blocks using graduated colors.

Join a table

US Census data is a reliable and helpful resource for any research project that involves looking at demographic data and is collected with geographic detail that ranges from fine to general granularity. Almost all the data is stored as values in tables and not geoenabled. However, you can perform several steps to turn this data into a map. First, you will view household income data provided by the US Census Bureau.

1. In ArcGIS Pro, open RetailSiteStudy.aprx from the GTKAGPro\RetailSiteStudy folder. The project opens a map of the city of Houston census block groups and a basemap.

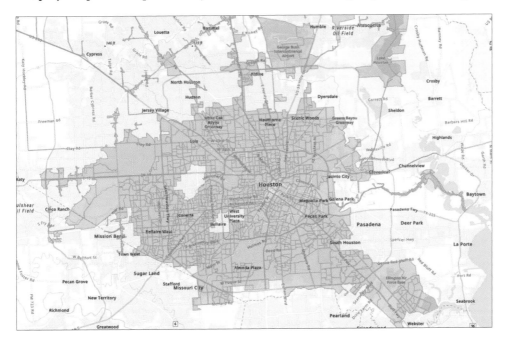

2. Add the median_household_income table to the project from GTKAGPro\RetailSiteStudy.gdb.

3. Open the median_household_income table to review it.

    The table includes the columns GEOID and Median_HHI. The GEOID name is a standard identifier that the US Census Bureau provides to identify census blocks (a division of a census tract). The Median_HHI column represents median household income (in US dollars). This will be one factor in your decision-making process to determine a potential site for a retail location. However, knowing only the household income is not so useful on its own. You will join it to the city of Houston census block layer to make the information spatial.

OBJECTID *	GEOID *	Median_HHI
1	15000US481576701011	45318
2	15000US481576701012	33022
3	15000US481576701013	46979
4	15000US481576701014	51755
5	15000US481576701021	35600

4. Close the table.

5. Open the Census Block Demographics attribute table to review it.

6. Join Census Block Demographics to median_household_income. Base the join on the GEOID field in both tables. You can ignore the index warning.

(Remember, you learned how to join tables in chapter 3.)

REMIND ME HOW
1. Go to the Census Block Demographics attribute table pane options menu, point to Joins and Relates, and click Add Join.
2. In the Geoprocessing pane, complete the details for the Add Join tool. For Input Join Field, click GEOID. Make sure the Join Table is set to median_household_income. The Output Join Field should automatically populate as GEOID.
3. Run the tool.

7. In the attribute table, scroll to the right to view the joined attributes from the median_household_income table.

Now you can see the median household income for each census block in the city of Houston in the Median_HHI column.

Symbolize using graduated colors

With median household income added for each census block, you will now visualize the median household income with graduated colors on the map. This will be one factor of consideration when doing the analysis for prospecting a retail site location. The median household income categories are as follows:

- Low Income: Less than $38,750
- Moderate Income: $38,751 to $63,250
- Moderately High Income: $63,251 to $ 97,800
- High Income: $97,801 to $156,950
- Very High Income: $156,951 to $250001

You will symbolize the data based on these categories.

1. Close any open tables. With the Census Block Demographics layer selected, go to the Feature Layer contextual tab, open the Symbology drop-down list, and click Graduated Colors.

   A random color scheme is applied to the census tracts. You will change the symbology settings to categorize the area median-income percentages.

2. In the Symbology pane, from the Field drop-down list, click Median_HHI, and from the Method drop-down list, click Manual Interval.
   The Manual Interval method allows you to create class breaks based on custom values.

3. Choose Orange-Red (5 Classes) from the Color Scheme list. Click the Show Names box to help you find the color scheme.

4. The Upper Value column will set the class breaks, so click the first field, type **38750**, and press Enter.

5. Continue to enter the upper values for the remaining categories:
   - Upper value: **63250**
   - Upper value: **97800**
   - Upper value: **156950**
   - Upper value: **250001**

6. In the Label field, set the labels for the categories.
    - **Low Income**
    - **Moderate Income**
    - **Moderately High Income**
    - **High Income**
    - **Very High Income**

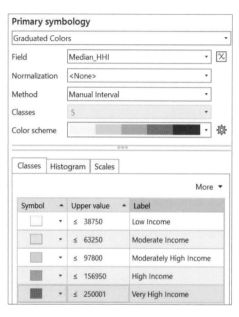

7. Click the More down arrow, and click Format All Symbols. Click Properties, and change the outline color to Soapstone Dust. Click Apply.

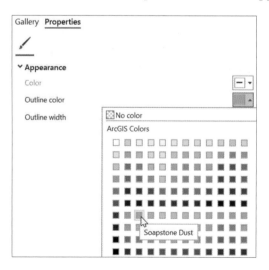

8. On the Feature Layer tab on the ribbon, in the Effects group, adjust the transparency from 50% to **0**%.

The layer originally had a 50% transparency so that labels on the basemap can still be seen.

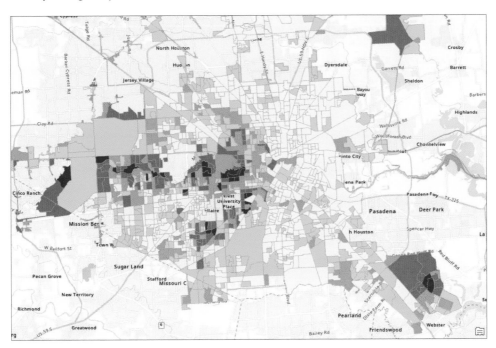

A color scheme is now applied so that you can visualize the areas that have differing levels of median household income. Moderate to very high-income areas are symbolized with gradually darker shades of orange-red and become emphasized. You can see areas on the west side of Houston and across a band through the center that have higher median household incomes that may support a large retail location (but may also be more expensive). Some census tracts may not have any symbology, which is often the case for areas designated as parks, airports, or port lands.

9. Save the project.

10. Keep the project open, and continue to exercise 6b.

This layer has been prepared for further analysis. Before you perform more geoprocessing, you will add more layers by geocoding data.

ON YOUR OWN

To learn more about data classification in ArcGIS Pro, go to ArcGIS Pro Help > Maps and Scenes > Layers > Layer Properties > Symbology > Symbolize Feature Layers > Data Classification Methods.

Exercise 6b: Geocode location data

Estimated time to complete: 30 minutes

An address allows someone to easily find a location of a business or home while walking or driving. A GIS requires similar information to produce the same result. A person must know two main pieces of information: the address of the business or home and knowledge of or reference to where that address is, such as street, main intersection, or neighborhood. Likewise, a GIS needs map or street information to know where to create address points. The process of creating map features from addresses, place-names, and similar information is called geocoding.

For your bicycle retail site location project, prospective retail locations are provided and must be geocoded—you must add a plain text file with addresses, names, and other information to your map. You will use a city of Houston street layer as the reference data—the spatial data that contains information about all the addresses. An address contains certain address elements (such as address number, street type, direction, and zip code) and comes in a range of formats. When you geocode the table of retail locations, you will use the reference data to create an address locator to create point features that represent the locations of the addresses (also known as locator styles). When geocoding your table of addresses, the address compared with the address locator.

Exercise workflow

- Create an address locator based on city of Houston streets.
- Geocode addresses of retail site prospects.
- Interactively rematch addresses.
- Geocode using the ArcGIS World Geocoding Service.

GEOCODE USING THE ARCGIS WORLD GEOCODING SERVICE

Obtaining accurate reference data can often be difficult. The ArcGIS World Geocoding Service references rich and high-quality point and street address data, places, and gazetteers that cover most of the world. The service is available for users who are signed into ArcGIS Online and operates under a credit-based use model.

Create an address locator

An address locator defines which reference data will be used to search for addresses, which elements of the reference data will be searched against, and what format the address data will use. ArcGIS Pro comes with predefined locator roles that you can use to create locators. A role includes the elements of the address that are present (such as street name, number, direction) and the elements that will be used to locate your addresses.

Each locator role is unique and has specific requirements regarding the reference data it can use to match the format of addresses being geocoded. Some locators have more require- ments than others, which allows for a wide range of accepted address formats. Using the appropriate role depends on knowing what type of reference data you have and what format your addresses will use. In this case, you will use a Street Address role.

1. Continue working with the map from exercise 6a. On the Analysis tab, click Tools.

2. In the Geocoding Tools toolbox, search for or browse to the Create Locator tool. Open the tool.

3. Select United States from the Country or Region drop-down list. Click the Browse button next to the Primary Tables, and click the roads layer in the RetailSiteStudy geo- database. Select Street Address from the Role drop-down list.

The roads layer acts as the reference data. Since you are using only one layer as the pri- mary table, it is assigned the role of Street Address. This layer has attributes for each street segment that indicate the range of addresses found on it. When you geocode an address, the address location is interpolated from the range of addresses found for a given segment.

The Field map may automatically populate with fields from the street layer that match the fields that the address locator style requires. Required fields are noted by an asterisk.

4. Under Field Mapping, set these fields as follows:
    - Left House Number From: LF_ADDR
    - Left House Number To: LT_ADDR
    - Right House Number From: RF_ADDR
    - Right House Number To: RT_ADDR
    - Street Name: NAME

5. Set these additional fields as follows:
    - Left City: CITY_L
    - Right City: CITY_R
    - Left State: STATE_L
    - Right State: STATE_R
    - Left ZIP: ZIP_L
    - Right ZIP: ZIP_R

6. Click the Browse button next to the Output Locator text box. Save the address locator as **Houston**_Locator to your RetailSiteStudy project folder. Set the Language Code as English.

Warning: Make sure you are not trying to save it to the geodatabase; address locators must reside in folders.

7. Click Run to create the locator.

    Creating the locator will take a few moments as ArcGIS Pro analyzes the street layer to create the address locator. You can view the progress at the bottom of the Geoprocessing pane. When the address locator is complete, it is a added to the project.

8. Dismiss the warning message.

9. In the Catalog pane, expand the Locators folder to verify that the locator was added.

    The other address locator—ArcGIS World Geocoding Service—is included with ArcGIS Pro and is configured to help you locate places on a map in situations in which you may not be able to create your own.

    A locator also determines other settings for geocoding, including what makes up a matched address, search parameters, and tolerance for spelling errors. You will see more of these settings next.

USING THE LOCATE PANE AND ADDRESS INSPECTOR TOOL

You can use the Locate pane and the Address Inspector tool (in the Inquiry group of the Map tab) to find addresses using a single locator or ArcGIS World Geocoding Service. This is useful for finding addresses one at a time or finding places for which you may not know the address, such as a business or public place. A pop-up appears that displays information for the address or place closest to the location clicked on the map. To learn more about using the Locate pane, consult ArcGIS Pro Help.

Geocode addresses

Now that the locator is ready, the next step is to geocode the list of prospective retail sites, which you'll find in your project folder. The output of the geocoding process can be either a shapefile or a geodatabase feature class of points. The geocoded data has all the attributes of the address table; some attributes of the reference data; and some additional attributes, such as the x,y coordinates of each point.

1. In the Catalog pane, expand the RetailSiteStudy folder, and add the retail_site_prospects.csv file to the current map. Open the table.

   The table contains many attribute columns, including the addresses of 28 retail site prospects, including full address, city, state, and zip code. It also includes square footage, purchase cost, and cost per square foot. You'll also notice some additional walk, transit, and bike ratings for each address—these ratings score how desirable a site is based on its accessibility by walking, taking public transit, or biking. The address information is added during the geocoding process.

2. Close the table, or move it out of the way.

3. In the Contents pane, right-click retail_site_prospects.csv and click Geocode Table. In the Geocode Table pane, read the guided workflow, scroll down, and click Go to Tool.

4. For Input Table, make sure retail_site_prospects.csv is selected.

5. Select HoustonLocator from the Input Locator drop-down list. If it's not available, click Browse next to the Input Locator box. Browse to Project > Folders > RetailSiteStudy, and select HoustonLocator or to Project > Locators > HoustonLocator.

6. Click the Browse button next to the Output Feature Class box. Browse to Project > Databases > RetailSiteStudy, and type **retail_site_prospects_Geocoded**. (The correct output name may already be there.) Click Save.

The Geocode Table settings are set and ready to run. Your display should look as follows:

7. Run the tool to geocode the retail site prospects.

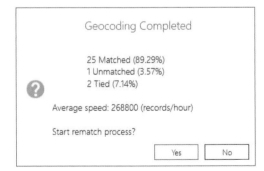

A dialog box appears that shows what addresses from the retail site prospects table are matched, unmatched, and tied with the roads data. All but three addresses are matched. You will perform an address rematch in the next exercise.

8. Click No to close the Geocoding Completed box.

You will get a chance to rematch them in the next exercise.

9. Zoom in so that you can see the retail site prospects better. (Turn the retail_site_pros-pects_Geocoded layer on and off to better see the point features.)

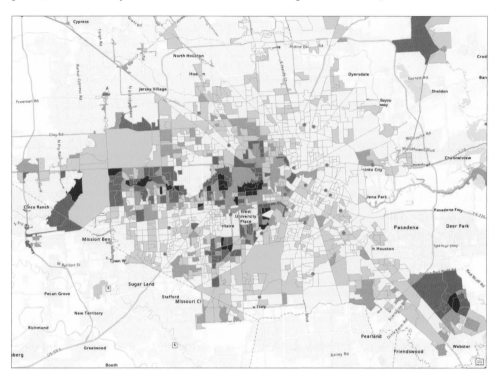

The retail site prospects feature class is added to the map when the geocoding is complete.

10. Open the retail_site_prospects_Geocoded attribute table.

The table contains 28 records, one for each retail site prospect that is in the retail_site_pros-pects.csv file. As a result of the geocoding, several additional attributes are added. The Status attribute column tells you whether each address is matched (M), unmatched (U), or tied (T). The Score attribute column tells you how well it scores when comparing the table with the reference data. The Match_type attribute shows how an address was matched—in this case, automatically (A) matched. The Match_addr attribute shows the actual address to which the retail site was geocoded, and the Side attribute tells you whether an address is on the right or left side of the street. The original fields of the table are preserved.

ON YOUR OWN

Sort the Retail Site Prospects table to see which addresses are matched, unmatched, or tied. Review the addresses to see why they may not have been geocoded. If you want, use the Locate pane to see if the ArcGIS World Geocoding Service can find the addresses you are reviewing. (Do not add points though. You will interactively rematch unmatched addresses next.)

Rematch addresses

It is good practice to review all geocoded addresses and go through the address rematching process to ensure that the geocoded addresses are correct. In this instance, several retail site prospects were not properly geocoded—it could be due to incorrect address data or addresses that were geocoded to the wrong location. You will use the Rematch Address feature to identify the unmatched addresses and look at the tied addresses to interactively rematch them.

1. Turn off the Census Block Demographics layer (the Topographic basemap allows you to see street names).

2. In the Contents pane, right-click the retail_site_prospects_Geocoded layer and click Data > Rematch Addresses to open the Rematch Addresses pane.

The Rematch Addresses pane allows you to review which addresses are unmatched or insufficiently matched, and then go through the parameters for each address and adjust them to make a match. When rematching addresses in this exercise, your view may change to look like the next image. To return to the original view of the points, in the Contents pane, right-click retail_site_prospects_Geocoded and click Zoom to Layer.

Locator	
Houston_Locator	
Unmatched Matched Tied	
Match_addr	
Address or Place	49000 Travis St
Address2	<Null>
Address3	<Null>
Neighborhood	<Null>
City	<Null>
County	<Null>
State	<Null>
ZIP	<Null>
ZIP4	<Null>
Country	<Null>
Auto Apply ◯	

The Rematch Addresses pane displays an address that is unmatched. Candidates are listed in the lower portion of the pane. However, the first address has no candidates.

3. Make sure that Auto Apply is turned on so that any changes in the pane are applied on the fly (which will speed up the candidate search).

You can use the arrows to browse through each address individually in any of the three categories: Unmatched, Matched, or Tied.

4. In the Rematch Addresses pane, for the first address, change Address or Place to **4900 Travis St**, and press Enter. Zoom out if needed to see the candidate location, which is highlighted in cyan on the map.

One candidate is found and is listed in the lower portion of the Rematch Address pane: Candidate A – 4900 Travis, Houston, TX, 77002, with a score of 90. You can click the candidate to show its location on the map (if it is not already highlighted).

If your Topographic basemap layer is not showing when zoomed in close, you can adjust its visibility. In the Contents pane, double-click the World Topographic Map layer. Change the In Beyond (maximum scale) value to 1:125, and click OK.

5. Make sure that candidate A is selected in the Rematch Addresses pane, and click Match (the green check mark) to match the address.

The Rematch Address pane is updated—the Unmatched category is no longer listed because all unmatched addresses are matched. Now there are only two tied addresses remaining. The retail_site_prospects table may not have matched because of missing information. You may have previously used the Locate pane to find addresses; if you have used it before, the next step should look familiar.

6. Click the Tied tab to review the tied candidates.

The first tied address is 2407 Southmore. It has two tied candidates, both with a score of 100. Candidate A—2407 Southmore, Houston, TX, 77004 and candidate B—2407 Southmore, Pasadena, TX (without a zip code). The green icon to the left of the candidate A icon indicates that it is currently mapped. Both candidates have the same address. However, they are in different locations on the map.

7. Click candidate B to see it in closer detail on the map.

Based on your knowledge with the prospects, you have been asked to look for the Southmore Blvd location. As you can see on the basemap, candidate B is on E Southmore Ave, so candidate A is the correct address to rematch to.

8. Make sure candidate B is selected and click the Unmatch button.

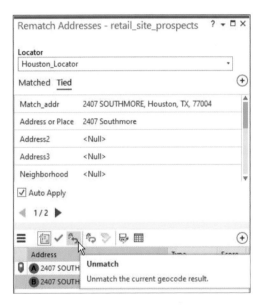

The Southmore candidates are unmatched, and the Unmatched tab appears again.

9. Click the Unmatched category. Make sure candidate A—2407 Southmore, Houston, TX, 77004—is selected and click the Match button.

10. Click Tied again.

Only one tied address remains: 1804 Main St. The top three candidates are the most likely ones based on the tied score. However, you will use additional information from another layer to help you decide which is the correct candidate.

11. On the Contents pane, make the Commercial Land Use layer visible. These are designated properties zoned for commercial use.

On the Rematch Addresses pane, click to view the top three candidates (A, B, and C). Select candidate B to view it on the map.

The site seems to indicate a commercial property (compared with the other top two candidates, which are not).

12. In the Rematch Addresses pane, look at the other field mapping values.

The existing zip code of 77999 doesn't match the candidate zip code of 77009.

13. Scroll to the ZIP field and change it to **77009**, and press Enter.

The candidate list refreshes and the score increases, creating a new top matching candidate.

14. Select the candidate A–1804 Main, Houston, TX, 77009, and click the Match button.

All the tied results are now matched. The Tied category in the Rematch Addresses pane disappears, and only the Matched category remains. Now, to ensure there aren't any other errors, click the left and right arrows to view the other matched addresses.

15. Click the right arrow through the matched addresses until you get to the seventh one, 5761 Cullenen Blvd.

You will notice that the Match_addr field in the upper portion does not match the Address or Place field shown in the Matched Address area. It is possible that the address in the retail_site_prospects table was incorrect. You will look up the address to verify it.

16. In the Locate pane, look up **5761 Cullenen, Houston, TX, 77021**.

The Locate pane is in the Inquiry group of the Map tab.

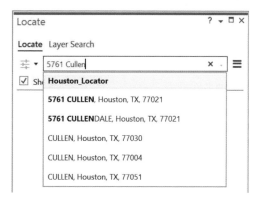

As you type the address, 5761 CULLEN appears as a result. The street name in the retail_ sites_prospects table was incorrectly spelled, but the locator has directed you to the correct address.

17. In the Rematch Addresses pane, change the Address or Place to **5761 Cullen Blvd**, and press Enter.

Candidate A matches the address in the Locate pane.

All the addresses are geocoded properly now. All the retail site prospects are visible on the map. In the next exercise, you will continue to use analysis tools to identify which of the sites will be most suitable based on other site selection criteria.

18. On the Edit tab, in the Manage Edits group, click Save. When prompted to save all edits, click Yes.

19. Save the project, close the attribute table, and close the Rematch Addresses and Locate panes.

20. Keep the project open, and continue to exercise 6c.

ON YOUR OWN

You can enter and look up addresses directly in ArcGIS Online. You can also add CSV files to any ArcGIS Online map (up to 1,000 features) as long as the file has supported fields, such as latitude, longitude, or address. Records are added to the map as points, which can then be shared with others or exported as data. To learn more about adding CSV files, consult ArcGIS Pro Help.

Exercise 6c: Use geoprocessing tools to analyze vector data

Estimated time to complete: 45 minutes

For this exercise, you will continue using analysis to identify retail sites that may be suitable based on a certain criterion. When looking for a suitable retail location, you'll examine existing bike lanes and proposed bike lanes in the city of Houston. You will look for retail site prospects within a specific range to bike lanes and select sites that are in a certain range of median household income. Then, you will look at additional criteria to help you narrow down which retail sites may be the most suitable.

Exercise workflow

- Create a buffer around existing and proposed bike lanes.
- Merge and dissolve bike lane buffers.
- Clip prospective retail sites to bike lane buffers.
- Select retail sites based on median household income by attribute and location.
- Perform a spatial join between the retail site buffers and bike stations.
- Analyze attributes and other criteria.

Create buffers

Buffers are polygons that are created around a feature at specified distances. To look for retail site prospects that are within a quarter mile of bike lanes, you'll first create buffers of existing and proposed bike lanes. These will be the criteria since many bike lanes are on main roads and highly trafficked routes, and sites within buffers should improve visibility and accessibility by customers whether they are driving, walking, or biking.

Although the Pairwise Buffer tool creates buffers of only one distance at a time, the Multiple Ring Buffer tool can create concentric buffer rings to represent multiple distances.

1. Continue working with RetailSiteStudy.aprx.

2. Hide all layers except Existing Bike Lanes, Proposed Bike Lanes, and the basemap, and zoom to the Existing Bike Lanes layer.

3. In the Geoprocessing pane, open the Pairwise Buffer tool (Analysis Tools > Pairwise Overlay).

    You will use the Pairwise Buffer tool to create a distance of a quarter mile around existing bike lanes. This measurement is a simplified representation of accessibility. Many more factors help define what accessibility means in real terms: mode of transportation, walkability, and road infrastructure.

4. For Input Features, click Existing Bike Lanes.

5. For Output Feature Class, type **ExistingBikeLanes_Buffer**.

6. In the Distance box, type **0.25**, and press Tab.

7. In the buffer unit box, click US Survey Miles.

8. For Method, keep Planar.

9. For Dissolve Type, click Dissolve All Output Features into a Single Feature.

10. Run the tool.

11. Create another buffer around Proposed Bike Lanes, also with a **0.25** distance, and name it **ProposedBikeLanes_Buffer**.

12. Zoom in to get a better look at the buffers.

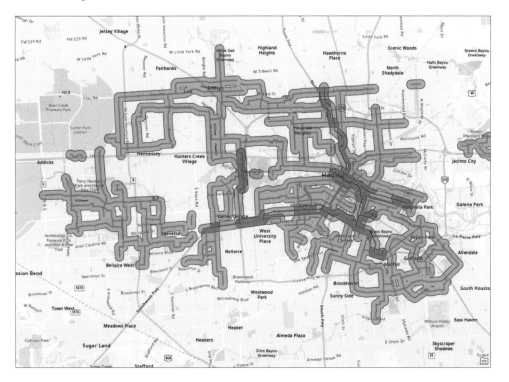

At a glance, you can see a visual cluster of bike lanes in the city of Houston core. You will *merge* these two buffers together and *dissolve* them to form one combined buffer.

Merge and dissolve features

The Merge tool merges multiple layers, and the Dissolve tool aggregates multiple features into one.

1. Open the Merge tool (Data Management > General).

2. For Input Datasets, add ExistingBikeLanes_Buffer and ProposedBikeLanes_Buffer. Name the output **BikeLaneBuffer_Merge**.

3. Click Run.

4. Open the Pairwise Dissolve tool (Analysis Tools > Pairwise Overlay).

5. For Input Features, add BikeLaneBuffer_Merge. Name the output **BikeLanesBuffer_Dissolve**.

6. Click Run.

    The single buffer layer will now be used as the clip features to extract retail site prospects.

Clip features

The Pairwise Clip tool extracts features using other features as a "cookie cutter" or trim boundary. This tool is particularly useful because you want to find only the retail site features that are within the bike lane buffer, essentially within a quarter mile of an existing or proposed bike lane.

1. Turn on the retail_site_prospects_Geocoded layer, and turn off BikeLaneBuffer_Merge, ProposedBikeLanes_Buffer, and ExistingBikeLanes_Buffer.

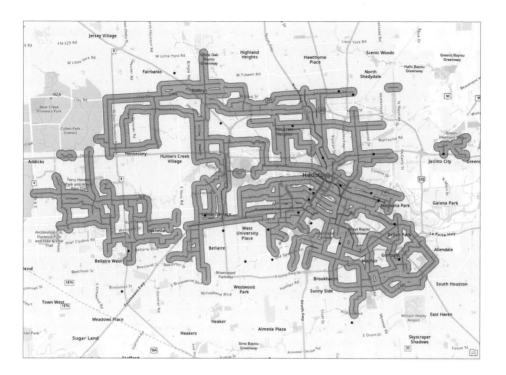

Reviewing this layer gives you an idea of what you are trying to extract—all the retail site prospects that fall within the boundary of the bike lane buffer.

ON YOUR OWN

Try using the Select tool to manually select retail site prospects. Learn how to use the Select options to isolate layers to make it easier to select features. When you finish, clear all selections.

2. Search for and open the Pairwise Clip tool (Analysis Tools > Pairwise Overlay).

3. In the tool pane, for Input Features, click retail_site_prospects_Geocoded.

4. For Clip Features, click BikeLanesBuffer_Dissolve.

5. For Output Feature Class, type **RetailSiteProspects_Clipped**.

6. Run the tool.
   The clipped retail site prospects layer is added to the map.

7. Turn off all layers except RetailSiteProspects_Clipped and BikeLanesBuffer_Dissolve.

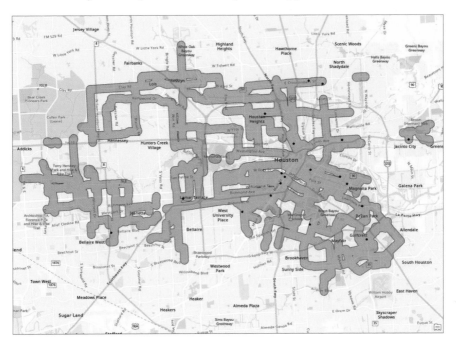

Now only the retail site prospects (assigned a random color) that are clipped to the bike lanes buffer layer are visible.

8. Open the attribute table of the RetailSiteProspects_Clipped layer.

The *clip* operation reduced the number of retail sites from 28 to 19. You will further reduce the number of retail sites by creating more selections based on additional criteria.

9. After examining the table, close it.

Select by attribute and location

To isolate the census tracts that you have seen throughout this project, you will use the Select By Attributes and Select By Location tools. The Select By Location tool can select features by calculating spatial relationships between layers. This step is also known as a *location query*.

1. In the Contents pane, make RetailSiteProspects_Clipped and Census Block Demographics the only visible layers.

   The high-end bicycle retail location will require you to look at who the optimal customer is in terms of demographics, based on a profile of your customer's economic situation. In this simplified example, you will look at moderately high to very high median house-hold income as an indicator. Customers in these areas may have additional disposable income to spend on high-end bicycles, maintenance, and service. Many questions can be addressed by looking at the prospective location's local economic data.

2. On the Map tab, in the Selection group, click Select By Attributes.
   - For Input Rows, click Census Block Demographics.
   - For Selection Type, click New Selection.
   - Create the expression: **Where Median_HHI is greater than 63250.**
   - Click OK.

3. On the Map tab, click Select By Location.
   - For Input Features, click RetailSiteProspects_Clipped.
   - For Relationship, click Within.
   - For Selecting Features, click Census Block Demographics.
   - For Selection Type, click Select Subset from the Current Selection.
   - Click OK.

   Selecting a subset from the current selection selects only the retail site prospects that are within the higher median income census block groups.

4. Open the RetailSiteProspects_Clipped attribute table. Click Show Selected Records to show only those features that are selected.
   Your selection criteria have narrowed the list to just seven retail site prospects.

5. Close the attribute table.

6. In the Contents pane, right-click RetailSiteProspects_Clipped, go to Selection, and click Make Layer from Selected Features.

7. Turn off all layers except RetailSiteProspects_Clipped Selection, Census Block Demographics, and World Topographic Basemap. Clear selected features.

Create a spatial join

Up to this point, you have used several criteria to narrow the remaining retail site prospects. You will continue to evaluate using additional criteria. The Bike Stations layer represents shared bikes that members of the community can rent for short rides. Combined with the bike lanes, the placement of bike stations also represents corridors of active cycling infrastructure and offers good visibility for a retail site. You will create a buffer around your final selection of retail site prospects and then perform a spatial join to include attributes of nearby bike stations.

1. Turn on the Bike Stations layer.

2. Prepare another buffer of **1** mile around the RetailSiteProspects_Clipped Selection layer. Name it **RetailSiteProspects_Buffer**. Run the tool.

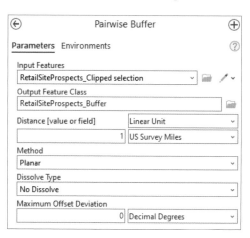

3. On the Analysis tab, in the Tools gallery, click Spatial Join.

4. In the Spatial Join pane, set these parameters:
   - For Target Features, click RetailSiteProspects_Buffer.
   - For Join Features, click Bike Stations.
   - Rename the Output Feature Class **Retail_Buffer_SpatialJoin**.
   - For Match Option, click Intersect.

5. Run the tool.

6. Turn off RetailSiteProspects_Buffer and Census Block Demographics.

7. Open the attribute table of the Retail_Buffer_SpatialJoin layer, right-click the Join_ Count attribute, and click Sort Descending.

8. Select the top three rows to highlight them (which also selects them on the map).

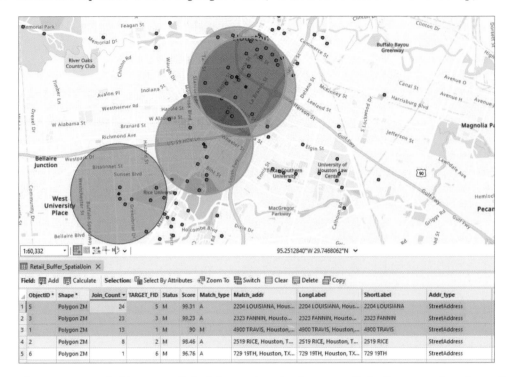

	ObjectID *	Shape *	Join_Count ▾	TARGET_FID	Status	Score	Match_type	Match_addr	LongLabel	ShortLabel	Addr_type
1	5	Polygon ZM	24	5	M	99.31	A	2204 LOUISIANA, Hous...	2204 LOUISIANA, Hous...	2204 LOUISIANA	StreetAddress
2	3	Polygon ZM	23	3	M	99.23	A	2323 FANNIN, Housto...	2323 FANNIN, Housto...	2323 FANNIN	StreetAddress
3	1	Polygon ZM	13	1	M	90	M	4900 TRAVIS, Houston,...	4900 TRAVIS, Houston,...	4900 TRAVIS	StreetAddress
4	2	Polygon ZM	8	2	M	98.46	A	2519 RICE, Houston, T...	2519 RICE, Houston, T...	2519 RICE	StreetAddress
5	6	Polygon ZM	1	6	M	96.76	A	729 19TH, Houston, TX...	729 19TH, Houston, TX...	729 19TH	StreetAddress

The Join_Count field counts all the bike stations in the one-mile buffer around each retail site prospect. Two sites have more than 20 stations around them, and the third site has 13 stations around it. These are good locations based on the amount of infrastructure in the area. Take note of the address of each location.

9. Clear the selection and close the table.

10. Right-click the Retail_Buffer_SpatialJoin layer and click Properties.

11. In the Layer Properties dialog box, click Definition Query and click New Definition Query. Create a query: **Where Join_Count is greater than or equal to 13**.

12. Click Apply and click OK.

13. Turn on the proposed and existing bike lane layers. Zoom in to get a better look.

14. To better distinguish the top three points in the RetailSiteProspects_Clipped Selection layer, turn the layer on and off.

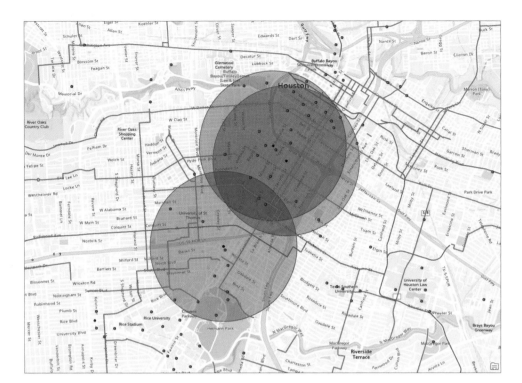

The remaining three retail site prospects are in a great location in the city core, in proximity to bike lanes and multiple bike stations, and in high median household income areas. Finally, you will look at a few other attributes that may help you decide which site is the most suitable. You will consider accessibility ratings for walking, transit, and biking and the square footage and cost of each retail site.

Accessibility ratings provide an index for the public to know how accessible a location is. Ratings higher than 80 provide a particularly good indicator of access. In this scenario, a bike rating higher than 80 means that cyclists can access the site easily by bike and can conduct their daily activities by bike.

In this exercise, cost and square footage are provided and simplified. Often, how much a location costs to purchase or rent could be the main decision factor.

15. On the Map tab, click the Explore button. Click any of the three potential locations and view the attributes in the pop-up. (Make the three points bigger if you can't see them.) Scroll down to the bottom of the pop-up and review the accessibility ratings, cost, and square footage.

Pop-up

▲ RetailSiteProspects_Clipped selection (1)
    `<empty>`

RetailSiteProspects_Clipped selection - `<empty>`

WALK	84
TRANSIT	66
BIKE	83
SQFT	9055
COST	1000000
COST_SQFT	110.44

◄ 1 of 1 ►   95.3869312°W 29.7305350°N

When comparing all three sites, the 4900 Travis St. site has the most square footage and has the lowest cost per square foot. It also has the best biking rating (83), but the lowest transit rating (66). The other retail sites have a slightly higher walking rating. As you can see, detailed information like this can factor into the decision-making process. In this scenario, all three sites are suitable candidates and worthy of consideration.

You have finished this chapter. If you are not continuing to chapter 7, exit ArcGIS Pro.

Summary

Using GIS can help establish a business location by analyzing all the data that is available to choose the best possible location. Whether the goal is to expand the current business or find a new franchise location, it is essential to use a fact-based approach. With layers of information about local geography, roads, land use, demographic data, socioeconomic conditions, and more, GIS can provide analysts, planners, and the community solutions to many challenges. It is also important to understand your customers, spending behaviors and patterns, and other trends. Although it is an advanced feature in ArcGIS, another way to do proximity analysis is to create trade areas—usually based on travel time or mode of transportation.

Glossary terms

geocoding
address locator
overlay
buffer
merge
dissolve
clip
location query

CHAPTER 7

Analyzing spatial and temporal patterns

Exercise objectives

7a: Create a kernel density map
- Select by attributes.
- Create a kernel density.

7b: Perform a hot spot analysis
- Run the Optimized Hot Spot Analysis tool.
- Create a space-time cube.
- Visualize a space-time cube.
- Run the Emerging Hot Spot Analysis tool.

7c: Explore the results in 3D
- Switch to a local view.
- Change 3D visualization styling.

7d: Animate the data
- Enable time.
- Animate using the Time slider.
- Animate using the Range slider.

Incidents of crime are neither uniform nor randomly organized in space and time. Crime mapping analysts can use GIS to identify spatial patterns to gain a better understanding of the role of location, proximity, and opportunity while providing key decision-makers with information to put crime prevention solutions in place.

Time is an important variable in analyzing crime. When using temporal data—in this case, crime incidents collected at a specific location and specific time—it is possible to see how patterns of crime shift over time and geography. If data is collected or surveyed to represent an instance in time—over the course of one day, one month, one year—the inclusion of time in a dataset's attributes gives temporal data the necessary information for a GIS to perform spatial and temporal analysis. The development of crime mapping has introduced the importance of location and time in criminology.

Scenario: You work as a crime analyst for the City of Philadelphia, and your role is to use GIS to examine criminal activity over time. You will do this by identifying crime hot spots and locating where those hot spots have gotten worse over time to help create spatially targeted crime prevention plans. Using a list of crime incidents that include *temporal data*—that is, data that has a time attribute—you will isolate a specific type of crime and create map layers that show crime locations as well as create crime hot spots.

GIS IN THE WORLD: SMART CRIME FIGHTING

The Ogden Police Department (OPD) Real Time Crime Center in Ogden, Utah, provides 24/7 support to law enforcement and uses ArcGIS to automatically link crime and other datasets maintained in different databases. OPD can perform advanced analysis and digitally map the results. These functions have allowed police staff to effectively deter crime and make arrests. Read more about the project in *ArcNews*, "Better Policing through Analysis": www.esri.com/news/arcnews/spring12articles/better-policing-through-analysis.html.

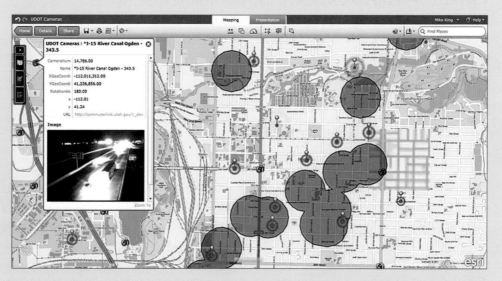

Figure 7.1. The Ogden Police Department relies on ArcGIS Server, as well as ArcGIS.com, to share information with the public and partner police agencies. Pictured: Known burglary suspects are shown inside a 1,000-foot buffer of forced-entry residential burglaries. Courtesy of Ogden (Utah) Police Department.

DATA

In the GTKAGPro\CrimeIncidents\CrimeIncidents geodatabase:

- crime—a feature class that represents crime incidents for the year 2014 in the City of Philadelphia (*Source: City of Philadelphia Police Department*)

Exercise 7a: Create a kernel density map

Estimated time to complete: 20 minutes

In this exercise, you will explore the crime layer of crime incidents and look more closely at its attributes. You will then create a kernel density layer to approximate crime hot spots to see where they occur in the city.

Exercise workflow

- Review data to see what makes it temporal.
- Select robberies by attribute, and create a layer from the selection.
- Create a kernel density layer to represent a crime hot spot.

Select by attributes

1. In ArcGIS Pro, open CrimeIncidents.aprx from your GTKAGPro\CrimeIncidents folder.

2. Open the crime layer attribute table.

    The crime layer contains crimes for the 2014 calendar year in the city of Philadelphia, Pennsylvania. The table lists 74,011 crime incidents, which are represented at a per block group address level, not as an actual address. This layer of anonymity protects the privacy of those involved with each incident. In addition, names and other identifiable information are not included with an incident. Notice the dispatch date and dispatch time attributes for each record. Although it is entirely possible to create a day-by-day temporal map with this level of data granularity, you will instead create a series of temporal maps for each month and use it in a time animation. In addition, you will focus on one type of crime: robberies, with and without a firearm. To perform this analysis, you will first extract data by crime type.

3. Close the attribute table.

4. On the Map tab, in the Selection group, click Select By Attributes.

5. In the pane, click crime for Input Rows. For Selection Type, click New Selection.

6. Under Expression, create this expression: **Where CRIME Is Equal To Robbery Firearm**.

7. Click Add Clause and create a second expression: **Or CRIME Is Equal To Robbery No Firearm**.

The expression includes the "or" *operator*, which selects both robbery attributes—with firearms and without firearms—and includes them in the new selection.

8. Run the tool.

Exactly 7,006 robbery points are selected on the map, indicated in the bottom right and representing robberies committed with and without firearms. Next, you will create a feature layer from the selection, in which you will work to extract robbery crimes by month.

9. Right-click the crime layer, and go to Data > Export Features.

10. For Input Features, click crime.

11. For Output Name, in the CrimeIncidents geodatabase path, type **robbery**.

12. Run the tool.

The robbery layer is now added to the map, and you can now select robberies by month.

13. Clear the selection and turn off the crime layer.

14. Create a selection by attribute in which DISPATCH_DATE is on or after 1/1/2014 12:00:00 AM and DISPATCH_DATE is on or before 1/31/2014 12:00:00 AM. Run the tool.

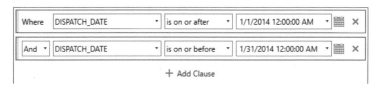

Exactly 640 robbery incidents during the month of January are selected on the map. Similarly, you will create a layer from this selection.

15. Export the selected features, and name the layer **robbery_jan**.

16. Clear the selection, turn off the robbery layer, and zoom to the robbery_jan layer.

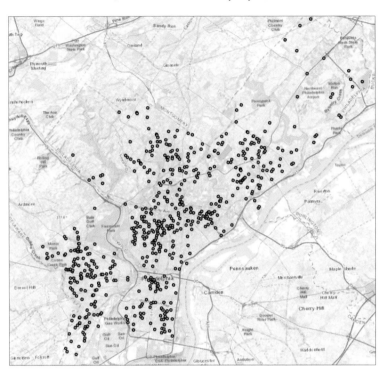

Next, you will create a kernel density for the robberies in January.

Create a kernel density

A *kernel density* calculates the density of features within an area around those features and is one of the most common techniques in crime mapping.

1. On the Analysis tab, in the Tools group, click Kernel Density (in the Analyze Patterns toolset).

2. In the Kernel Density pane, set these parameters:
   - For Input Point or Polyline Features, click robbery_jan.
   - For Population Field, make sure NONE is selected.
   - For Output Raster, replace the name with **robbery_jan_density**.
   - If necessary, change Area Units to Square Miles.
   - Retain the other default parameters.

3. Run the tool.

The kernel density is calculated and added to the map. The density is automatically classified with 10 classes and styled using a random color ramp (your colors may look different from the figure).

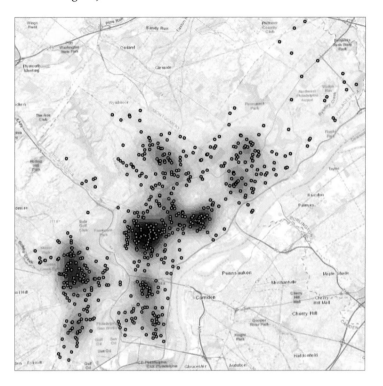

4. On the Raster Layer contextual tab, change the robbery_jan_density layer transparency to **40%**.

The least-dense class of the kernel density now has practically no color, and the remaining classes at 40 percent transparency allow you to see the basemap.

Next, you will use more analysis tools on the robbery layer to help visualize more of the hot spots.

ON YOUR OWN

Try creating a kernel density for other months to compare how each month differs. Try creating a kernel density for other categories of crime. Do some areas have more incidents of certain crimes than others? It is important to note that when comparing density surfaces, the classifications should be the same. Using different classifications causes different results.

5. Save the project.

Exercise 7b: Perform a hot spot analysis

Estimated time to complete: 45 minutes

You created a kernel density in exercise 7a to briefly see where areas of high density may occur. You will expand on how to build an analysis by using other types of analysis tools. In this exercise, you will run the Optimized Hot Spot Analysis tool to find statistically significant hot spots and cold spots of robberies. You will also create a space-time cube, a data pattern mining tool, to look for trends over time.

Exercise workflow

- Run the Optimized Hot Spot Analysis tool to create a hot spot map.
- Run the Create Space Time Cube tool to mine for data patterns.
- Visualize the space-time cube in 3D.
- Run the Emerging Hot Spot Analysis tool to see hot spot trends in the robbery data.

Run the Optimized Hot Spot Analysis tool

1. Continue working from the map in exercise 7a.

2. In the Contents pane, turn off the robbery_jan layer and the robbery_jan_density layer. Turn on the robbery layer.

3. On the Analysis tab, open Optimized Hot Spot Analysis.

4. In the pane, for Input Features, click robbery. Rename Output Features **robbery_OHSA**.

5. Retain the other parameters and run the tool.

6. Turn off the robbery layer.

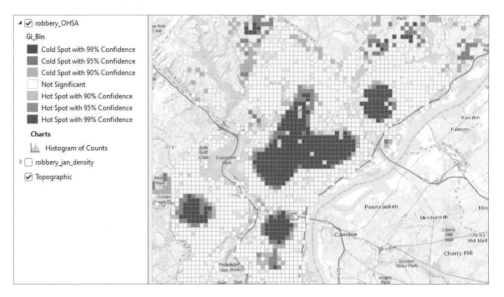

The optimized *hot spot analysis* performed on the robbery data is added to the map. The tool aggregated robbery incidents into weighted features and analyzed their distribution to identify an appropriate scale of analysis. It then determined which areas are significant, including areas that are hot (large clusters symbolized in red) and cold (smaller clusters symbolized in blue). Areas that are not significant are symbolized in white.

7. Open the robbery_OHSA attribute table.

Before you look at the values, it is important to make some assumptions used in statistics. You are essentially asking the questions: "Is this pattern the result of random spatial processes?" and "How likely is it that this pattern is random?" The initial assumption in this cluster analysis is that crime clusters happen in specific areas of the city and are not random. When it is *very* unlikely that the pattern is random, you reject the *null hypothesis*—which, in this case, is complete spatial randomness—and you can feel confident that the patterns you are seeing are evidence of underlying spatial processes. In other words, you have found statistically significant clusters of both high and low values that you can trust, and you can begin to ask questions as to why they occur, thereby rejecting the null hypothesis.

The intensity of robbery clustering is determined by the z-score returned in the output. Z-scores are standard deviations and are represented in the table as the attribute GiZscore Fixed 3987. As an example, a z-score of +2.5 means that the result is 2.5 standard deviations and is at a high confidence level (greater than 95 percent), typically indicating statistically significant clustering.

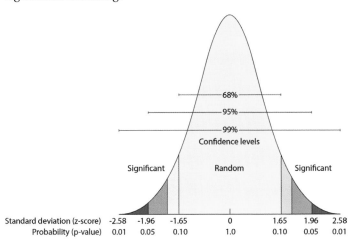

A probability graph shows confidence levels and standard deviations and probability values that indicate either the significance or the randomness of phenomena.

8. Use the Explore tool to select one cell in a red hot spot.

If you do not see a pop-up for the cell you selected, click the Explore tool down arrow and then click Visible Layers.

Pop-up   ∨ ☐ ✕
▲ robbery_OHSA (1)
    1116

robbery_OHSA - 1116

OBJECTID	1116
SOURCE_ID	1116
Counts	3
Shape_Length	3715.999786
Shape_Area	863040.900509
GiZScore Fixed 3987	6.949083
GiPValue Fixed 3987	0
NNeighbors Fixed 3987	61
Gi_Bin Fixed 3987_FDR	3

Most likely, the z-score of the selected cell is greater than +2.5. However, the p-value is a very low number, close to zero. The p-value represents the probability that the observed spatial pattern was created randomly. A very small p-value means that it is highly unlikely that the observed spatial pattern is the result of random processes. This small p-value can be an indication that robbery incidents are not entirely random and that many contributing factors can lead to their occurrence, such as socioeconomic factors in low-income neighborhoods.

9. Close the attribute table and pop-up, and clear any selection.

Next, you will create a space-time cube to add a temporal component to the analysis.

Create a space-time cube

The Create Space Time Cube tool summarizes the set of robbery incidents into a data structure by aggregating them into space-time bins. These bins can be viewed in 2D or 3D. Each bin contains the robbery points for a certain month, and the pattern of incidents over time is analyzed for all locations. Creating a *space-time cube* can help you aggregate the robbery points, and visualizing it will help you understand how the robbery data is distributed over a geographic area. This tool only creates the space-time cube—it does not visualize it. You will visualize the space-time cube immediately after creating it.

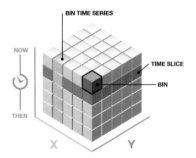

A set of points can be summarized into a data structure by aggregating them into space-time bins that form a cube.

1. In the Geoprocessing pane, search for and open **the Create Space Time Cube by Aggregating Points** tool.

2. For Input Features, click robbery. For Output Space Time Cube, type **robbery_STC.nc**. (Hint: It is important to include the .nc file name extension.)

The .nc file extension stands for the netCDF format (network Common Data Form) and is primarily used for storing multidimensional data such as temperature, but it can also be used for other data types. ArcGIS Pro uses the capabilities of the netCDF format and can display data in multiple dimensions, including time.

3. For Time Field, click DISPATCH_DATE. In the Time Step Interval box, type **1**. In the adjacent box, click Months.

4. Run the tool. When the tool is finished running, click View Details to see more about what was created. Click the Parameters tab if it is not already showing. When you have finished viewing the details, close the box.

The message provides insight into what was created, including the number of locations, the geographic size of each location, the total geographic size, and the number of time intervals. You will use another tool to visualize the space-time cube.

Visualize a space-time cube

The space-time cube you created for robbery incidents will be rendered in 3D, so you will create a 3D scene.

1. On the Insert tab, click the New Map down arrow, and then click New Local Scene.

2. In the Contents pane, under the Elevation Surfaces layer, expand Ground, and remove the default WorldElevation3D/Terrain3D (right-click and click Remove).

   Instead of elevation, time is being used as the vertical axis in visualizing the robbery space-time cube. By removing the service that provides elevation data, all the time values will start on the ground and at the same base, with an elevation equal to zero.

3. In the Geoprocessing pane, search for and open the **Visualize Space Time Cube in 3D** tool.

4. For Input Space Time Cube, browse to the robbery_STC.nc file, select it, click OK, and change these parameters in the tool pane:
   - For Cube Variable, click COUNT.
   - For Display Theme, click Value.
   - For the Output Features name, type **robbery_STC_3D**.

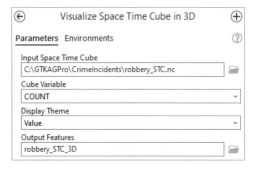

   The Count display variable shows the density of the robbery counts in each space-time bin.

5. Run the tool.

   After the tool is finished running, it may take a few moments to render the scene.

6. Zoom to the new robbery_STC_3D layer.

7. On the Map tab, click Explore. While holding down the mouse wheel button, drag the mouse to navigate the map and view the space-time cube.

The map shows space-time bins and the counts for robbery incidents. When zoomed in close enough to the scene, you can see 12 intervals, each representing a month, for each location (as shown in the next image).

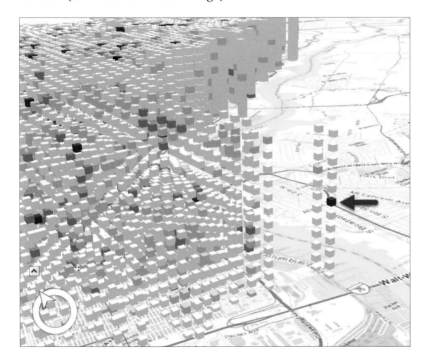

The legend in the Contents pane for the robbery_STC_3D layer shows that the darker intervals show higher counts. Although looking at the space-time cube alone provides some hints on the clustering of robberies, it is not conclusive. You will use another tool that specifically identifies trends in the clustering of space-time point data for a more detailed perspective.

Run the Emerging Hot Spot Analysis tool

The Emerging Hot Spot Analysis tool identifies hot spot and cold spot trends in your data. It essentially provides a hot spot analysis like the one at the start of this chapter.

1. Return to the 2D map, and turn off the robbery_OHSA layer.

2. In the Geoprocessing pane, search for and open the **Emerging Hot Spot Analysis** tool.

3. For Input Space Time Cube, browse to the robbery_STC.nc file, select it, click OK, and change these parameters in the tool pane:
   - For Analysis Variable, click COUNT.
   - For Output Features, type **robbery_STC_EHSA** as the new name.

4. Run the tool.

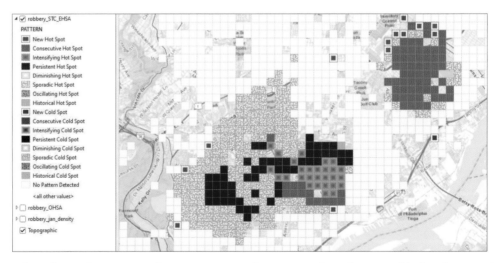

When the analysis is complete, an emerging hot spot analysis layer is added to the map. In the Contents pane, notice the many patterns listed to categorize the locations. Remember that the Emerging Hot Spots Analysis tool accounts for space and time and analyzes all robbery incidents to determine whether clustering is statistically significant.

5. Zoom in to the large hot spot at the center of the map.

6. Click a location that is classified as Persistent Hot Spot and review the pop-up.

In this classification, a persistent hot spot is a location that has been a statistically signifi-cant hot spot for 100 percent of the time-step intervals with no discernible trend indicating an increase or decrease in the intensity of clustering over time. This hot spot classification means that robberies in this area are frequent and concentrated.

7. Click a location that is classified as Intensifying Hot Spot and review the pop-up.

An intensifying hot spot is a location that has been a statistically significant hot spot for 90 percent of the time-step intervals. It is an area in which there has been an increase in the intensity of clustering over time and is statistically significant. The spatial correlation between intensifying hot spots and persistent hot spots is not random. Crime can have a spreading trend, which may be what is happening here.

8. Pan south on the map, click a location that is classified as New Hot Spot, and review the pop-up.

This area has many locations classified as new hot spots. A new hot spot is a location that is a statistically significant hot spot for the final time step and has never been a statisti-cally significant hot spot before. Although many factors can cause this area to become a significant cluster, the map can be used to help determine what is causing the increase in robberies during this time interval.

To learn more about the Emerging Hot Spot Analysis patterns, go to ArcGIS Pro Resources > Tool Reference > Geoprocessing Tools > Space Time Pattern Mining Toolbox > Space Time Pattern Mining Concepts > How Emerging Hot Spot Analysis Works.

9. Clear any selections you have made.

10. Make Scene the active map.

11. Run the Visualize Space Time Cube in 3D tool again.

12. For Display Theme, click Hot and Cold Spot Results.

13. Rename the output **robbery_STC_3D_Hotspots** and run the tool.

14. Turn off the robbery_STC_3D layer.

A 3D version of the results of the Emerging Hot Spot Analysis is added to the Scene map.

15. Save the project.

Exercise 7c: Explore the results in 3D

Estimated time to complete: 30 minutes

The ability to animate data is a great visualization tool to show changes over time. In exercises 7a and 7b, you completed the creation of crime hot spots for the City of Philadelphia. In this exercise, you will explore analytical results in 3D to observe patterns or trends that may emerge with the passage of time.

Exercise workflow

- Switch to a more scientific visualization display.
- Investigate hot spots and change visualization styling.

Switch to a local view

Depending on your ArcGIS Pro settings, the 3D view can be in global viewing mode, which means that the data is being reprojected to WGS84. For this dataset, with a projected coordinate system and a limited area of interest, you should work directly in its native coordinate system, NAD 1983 State Plane Pennsylvania South.

1. Continue working from the map in exercise 7b. Click Scene to make it the active map.

2. Turn off the robbery_STC_3D_Hotspots layer, and make the robbery_STC_3D layer visible.

3. In the Contents pane, right-click Scene and click Properties.

4. Click the Coordinate Systems tab to view the coordinate system details, expand Layers, and make sure that NAD 1983 StatePlane Pennsylvania South FIPS 3702 (US Feet) is highlighted.

   The local view is used for areas of smaller geographic extent in a projected coordinate system or when representing the earth's curvature is not required. In this case, the local view is using the NAD 1983 StatePlane Pennsylvania South FIPS 3702 (US Feet) coordinate system. Next, you will clip the extent.

5. On the Clip Layers tab, click Clip to a Custom Extent.

6. Under Get Extent From, click robbery_STC_3D, and click Apply.

   The custom extent is calculated from the robbery_STC_3D layer. Clipping the extent of the view will help isolate the area of interest.

7. Click OK.

The map properties are applied, and the scene view is updated. You will add a new field so that you can assign it the appropriate time.

8. Open the attribute table of the robbery_STC_3D layer, go to the options menu, and click Fields View.

☑ Visible	■ Read Only	Field Name	Alias	Data Type
☑	☑	OBJECTID	OBJECTID	Object ID
☑	☐	Shape	Shape	Geometry
☑	☐	TIME_STEP	Time Step ID	Long
☑	☐	START_DATE	Start Date	Date
☑	☐	END_DATE	End Date	Date
☑	☐	TIME_EXAG	Time Step ID Exaggeration	Double
☑	☐	ELEMENT	Element	Long
☑	☐	LOCATION	Location ID	Long
☑	☐	VALUE	COUNT	Double

The robbery hot spots are now all designated by time. Next, you will enable time so that the hot spots can be included in animations.

9. Close the Fields View and the attribute table.

Change 3D visualization styling

Using the standard or default symbology, the cubes will "separate" as you get closer. This type of visualization is fine for some use cases—for example, it helps with column-based understanding of the data and helps reveal any patterns that might be "wrapped inside" other items.

1. Turn off the robbery_STC_3D layer, and make the robbery_STC_3D_Hotspots layer visible.

2. Zoom in until you can see the time intervals separate.

Although this type of visualization works in some cases, in this mode it is harder to understand the spatial coverage of the analysis. In this case, you can resymbolize the content so that the cubes are in fixed real-world size, as flattened cubes that cover the spatial extent they represent.

3. Open the properties of the robbery_STC_3D_Hotspots layer.

4. On the Display tab, click the Display 3D Symbols in Real-World Units check box. Click OK.

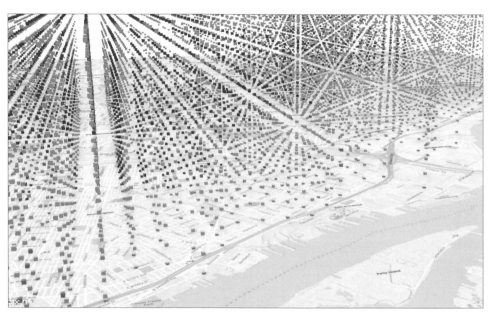

3D symbols displayed in real-world units display a constant measurable size, regardless of view distance. Next, you will change the symbology of the locations to make them real-world size so that they fit the physical space they represent. You can make this change by manually changing each symbol or using attribute-driven symbol properties (if available in the feature class), or you can use the attribute-driven property settings with constant values. You will use this third option because it is easy to access from the Symbology pane and will not change the legend in the Contents pane.

5. Right-click the robbery_STC_3D_Hotspots layer and click Symbology.

6. In the Symbology pane, click the Vary Symbology by Attribute button.

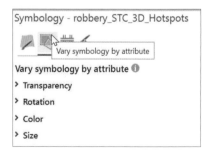

7. Expand the Size setting.

8. Clear the Maintain Aspect Ratio check box.

9. Next to the Height field, click the Expression Builder button.

10. In the Expression Builder window, type **50** in the text box, and click OK.

11. For Width, in the Expression Builder, type **250** in the text box, and click OK.

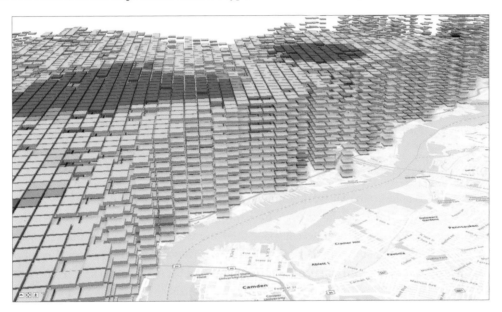

The hot spots of the space-time cube adjust according to the values you entered in the Size property of the Symbology pane. Next, you will adjust the colors to emphasize the significant hot spots. You can hide the insignificant locations by making them transparent.

12. In the Symbology pane, click the Primary Symbology button to return to the main symbology settings.

13. In the table of symbols on the Classes tab, click the symbol for Not Significant.

14. Click Properties. Click the Color down arrow, click No Color (at the top of the palette), and click Apply.

For the hot spots that are designated as 90% Confidence, you will make them partially transparent and therefore less prominent.

15. Click the back arrow and click the symbol for Hot Spot – 90% Confidence.

16. Click the Color down arrow and click Color Properties. In the Color Editor, change the transparency to **40%**, and click OK.

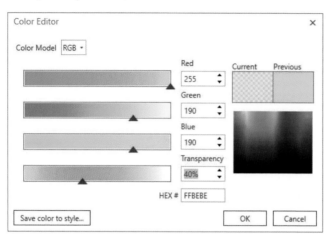

17. Click Apply.

The map is updated to reflect the changes in the Symbology pane.

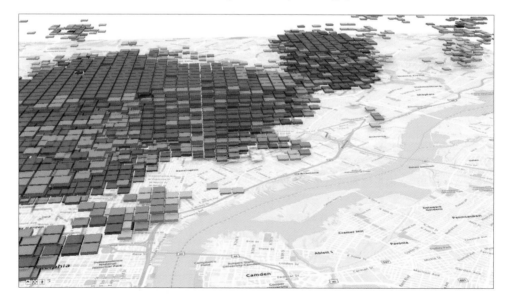

ON YOUR OWN

If you are having trouble seeing the hot spots, change the visibility of the robbery incidents layer—by either turning it off or adjusting its transparency. You can also enable the Swipe tool to reveal layers beneath the chosen layer. To use the Swipe tool, select the layer in the Contents pane, and click the Swipe tool on the Feature Layer contextual tab.

18. Save the project.

Exercise 7d: Animate the data

Estimated time to complete: 20 minutes

Exercise workflow

- Enable time for the 3D hot spot layer.
- Set up the Time slider and animate the data.
- Set up the Range slider to filter the data.

Enable time

The robbery incidents layer includes a DISPATCH_DATE attribute that you can use to indicate time. First, you will enable time for the 3D robbery hot spot layer.

1. Continue working from the map in exercise 7c. Make sure Scene is the active map.

2. Open the properties of the robbery_STC_3D_Hotspots layer.

3. On the Time tab, click Filter Layer Content Based on Attribute Values.

4. For Layer Time, click Each Feature Has Start and End Time Fields.

5. For Start Time Field, click Start Date.

6. For End Time Field, click End Date.

7. If the Time Extent does not refresh automatically, click Calculate, and click OK.

Filter using time
Use the map's time extent to filter the display of content for this layer

○ No time - content is always shown

○ The entire layer is shown within a fixed time extent

● Filter layer content based on attribute values

Layer Time	Each feature has start and end time fields ▾
Start Time Field	Start Date ▾
End Time Field	End Date ▾
Time Extent	12/31/2013 12:00:01 AM ▦ - 12/31/2014 12:00:00 AM ▦
	Calculate
	☐ Data is a live feed. Refresh rate is on the General tab.
Time Interval	● No pre-defined time interval
	○ View using a regular time interval
	Step [10] Seconds ▾
	○ View using unique times within the data

The Calculate button ensures that the full time extent—start time and end time—is specified. The extent can also be changed later when adjusting the time settings.

You may notice a bar appear near the top of the map or a small box in the upper-right
corner of the map that says Time. This Time slider indicates that the map has successfully
detected the temporal data that you specified.

Animate using the Time slider

1. If necessary, click the Time down arrow to expand the animation controls or point to
   the animation controls to see them.

2. On the ribbon, click the Time tab to make it active, and click the Enable Time button on
   the ribbon (View group) to enable it.

The Time tab contains all the settings to control animations. The time extent is indicated in
the Full Extent group showing a start time of 12/31/2013 and an end time of 12/31/2014.
Because the time extent shows a full 12 months, all the robbery incidents displayed represent a full year and are shown on the map. You will adjust the start time slightly.

3. In the Full Extent group on the ribbon, change the Start date to **01/01/2014**.

    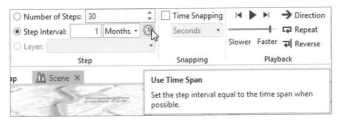

    Next, you will adjust the current time to show only the month of January 2014.

4. In the Current Time group on the ribbon, change the Start date to **01/01/2014** and the End date to **01/31/2014**.

    Setting the current time creates a map time that represents the month of January. On the map, notice that the number of robbery hot spots is reduced. (You may need to zoom out or pan to notice.) These robberies are the ones that occurred only in the month of January. The robbery points for other months are hidden until the map time is changed.

5. In the Step group, click the Use Time Span button to add values for the Step Interval option. Specify an interval of **1 Months**.

6. In the Playback group, click the Step Forward button.

    The map time moves forward one month, and the robbery hot spots for February are shown. On the Time slider and in the Current Time group, the February start and end times are displayed (2/1/2014 to 2/28/2014).

7.   In the Playback group on the ribbon, use the Speed slider to adjust the speed of the animation. Click Repeat to enable it.

8.   Click the Play All Steps button. Pan the map to view the animation over a wider area.

The animation begins to play and cycle through each month automatically.

9.   Click the Pause button at any time to stop the animation. Use the Step Back and Step Forward buttons to move to the previous or next interval, respectively.

These controls allow you to move map time month by month so that you can see the difference across each month.

Unique temporal slices are included in the visualization cache so that the animation displays and performs better on second viewing.

ON YOUR OWN
Experiment with the Time slider and adjust the playback options.

Animate using the Range slider

First, you will enable the range of values from an attribute. In this case, you will use the Hot Spot Analysis Bin (Count) attribute, which is the confidence level indicated in the legend. It ranges from −3 to +3, with zero classified as Not Significant.

1. Open the properties of the robbery_STC_3D_Hotspots layer.

2. On the Range tab, click Add Range.

3. For Start Field, click Hot Spot Analysis bin (COUNT).

4. For End Field, click Hot Spot Analysis bin (COUNT), and click Add.

5. Click OK.

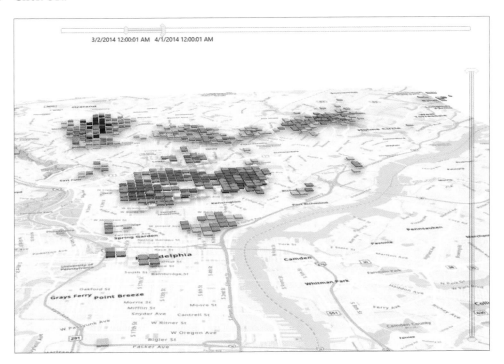

The Range slider functions similarly to the Time slider, except that instead of time as the variable, it uses numerical values of any attribute that you are interested in. You will isolate only the highest confidence hot spots by using the values specified in the Emerging_Count_ Hs_Bin attribute.

6. On the Range tab, in the Full Extent group, enter a Min Value of **2** and a Max Value of **3**.

The scene shows hot spots that are now filtered by time *and* range. Using this combination of time and range, you can provide information for specific instances in time. Isolating hot spot clusters can provide a narrative on when and where crimes happened.

The Range slider works in 2D and 3D, so you can use the same technique to easily filter the content you are viewing in a 2D map view as well.

ON YOUR OWN

Explore connecting the Range slider to other numerical properties, such as Location ID.

7. Save the project.

Summary

You have learned what temporal data is, how to work with it, how to visualize it in 3D, and how to create an animation in ArcGIS Pro. The strength of GIS lies in the ability to recognize patterns and the organic behavior of crime—particularly in clusters over time. Although crime mapping is only one part of an often complex crime reduction plan, hot spot maps are useful in providing a spatial sense of the nature of high-crime areas and where crimes are concentrated. However, further analysis is required to understand why certain crimes happen more often in certain areas. The evidence of crime clusters can influence key decision-makers in crafting effective crime reduction tactics and can help crime analysts use GIS to convey that information.

Glossary terms

temporal data
operator
kernel density
hot spot analysis
null hypothesis
space-time cube

CHAPTER 8:
Determining suitability

Exercise objectives

8a: Prepare project data
- Convert a line feature to a polygon feature.
- Clip raster layers.
- Merge rasters.

8b: Derive new surfaces
- Derive an aspect surface.
- Derive a slope surface.
- Derive a hillshade surface.
- Visually compare analysis outputs.

8c: Create a weighted suitability model
- Reclassify criteria rasters.
- Combine criteria rasters.

In this chapter, you will solve a common GIS problem: determining which areas are most suitable for a specific land-use purpose. To do so, you will take advantage of the raster data model.

As discussed in chapter 1, the first thing you should understand about a raster is that it is composed of a grid of *cells*, instead of discrete x,y coordinates, that define geographic entities. The cells contain values that are used to record and define geographic phenomena on the surface of the earth. Each raster cell represents a portion of the earth, such as a square meter or square mile, and usually has an attribute value such as elevation, soil type, or vegetation class associated with it. A surface, which is made up of raster cells, can be *discrete data*, which means that it shows distinct and discernible regions on a map, such as soil types, or it can be *continuous data*, which means that there are smooth transitions between variations in the range of data. Elevation data is one of the most common uses of a continuous raster. Typical raster datasets are extremely detailed and store a large amount of information, recording information about each cell.

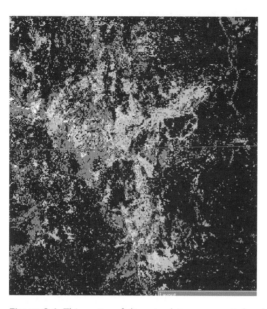

Figure 8.1. This raster of discrete data represents land cover. Data courtesy of Lake Tahoe Data Clearinghouse, US Geological Survey.

Figure 8.2. This raster of continuous data represents elevation. Data courtesy of the Office of Geographic Information (MassGIS), and the Commonwealth of Massachusetts Information Technology Division.

Map algebra—a language that combines GIS layers—is fundamental to raster analysis. You can evaluate raster cell values using functions and mathematical, logical, or Boolean operators. An output raster is the result of a cell-by-cell function performed on two input rasters.

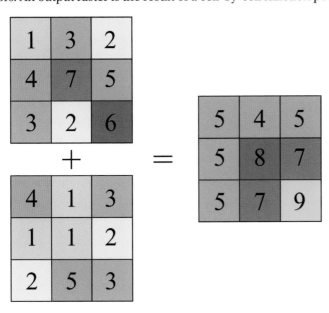

Figure 8.3. A 3 × 3 grid added to another 3 × 3 grid produces a result grid, in which the cell values represent the total of the inputs for each cell.

Every raster cell has a value. Some cells may have a value of zero (for instance, representing no precipitation), and some cells may have a value of *NoData*, which means that no values were recorded for that cell. NoData cells are ignored in raster calculations.

The ArcGIS Spatial Analyst extension contains several hundred geoprocessing tools to perform comprehensive, raster-based spatial analysis. These tools allow you to perform basic mathematical and logical operations, as well as raster dataset creation and processing.

In this chapter, you will work with both vector and raster data, and you will use ArcGIS Pro standard tools as well as Spatial Analyst tools.

If you are using student trial software, the Spatial Analyst extension is automatically licensed.

Scenario: Suppose you are a vineyard manager. Your employer owns a large property in Monterey County, California. Much of it is already planted with grapes, but there are several unused blocks of land. You will use your enological knowledge and GIS skills to select the most suitable block for planting additional vines.

To grow wine grapes in a moderate coastal climate such as in Monterey County, sunlight is the key ingredient—the more, the better, for rich fruit and high yield. Your specific criteria are that the land must be mostly south facing, for the longest sun exposure; the slope must be less than 14 percent, because you want the vineyard to be machine-harvestable; and finally, you want as little shade as possible. (For example, even if a block is south facing, it can have high slopes on either side, creating undesirable shade in the morning and afternoon.)

With your understanding of topography and GIS, you know that you must create and compare three layers: aspect, slope, and hillshade. You will have to do some preprocessing before you dive into the analysis. Once you have analyzed the property, you will overlay a feature class showing existing vineyards so that you can assess the available areas and decide which one is most suitable for planting.

Note: A special thanks to Tyler Scheid, Jonathan Vevoda, and Greg Gonzales at Scheid Vineyards for allowing us to use their data and general workflow. (Check out the fruit of their GIS-enabled efforts at www.scheidvineyards.com.)

DATA

In GTKAGPro\Vineyard\Vineyard.gdb:

- ned_1 and ned_2—elevation rasters (digital elevation models, or DEMs) for the study area (*Source: National Elevation Dataset from USGS*)
- planting_sites—a polygon feature class that outlines available planting areas (*Source: Scheid Vineyards*)
- property_boundary—a line feature class that delineates the property boundary of Scheid vineyards in Monterey (*Source: Scheid Vineyards*)
- Soils_clip—a soils survey layer that is clipped to the extent of the property_boundary feature class (*Source: USGS Soils Survey*)
- vineyard_blocks—a polygon feature class that shows where vineyards currently exist (*Source: Scheid Vineyards*)

To get the data in its current state, we downloaded elevation rasters (DEMs) from https://viewer.nationalmap.gov/viewer. The property boundary coincides with two large DEM files, named ned19_n36x25_w121x00_ca_centralcoast_2010 and ned19_n36x25_w121x25_ca_centralcoast_2010. Once retrieved, the rasters' extents were then reduced to minimize student download size. (The extents were minimized by hand-digitizing a rough study area polygon around the property boundary, and then using the Extract By Mask tool to clip a portion of the full-sized rasters. The study area was used as a *mask*—a means of identifying areas to be included in a geoprocessing operation.) You will perform a similar operation in exercise 8a.

Exercise 8a: Prepare project data

Estimated time to complete: 20 minutes

In this exercise, you will prepare data for a vineyard suitability analysis.

Exercise workflow

- Convert the line property boundary feature to a polygon feature so that it can be used as an extraction mask.
- Using the Extract by Mask tool, reduce the extents of the two elevation rasters so that they are clipped to the extent of the property boundary.
- Append one extracted raster to the other, resulting in a single raster that is coincident with the property boundary.
- Project the elevation raster so that it is in the same coordinate system as the property boundary.

Convert a line feature to a polygon feature

1. In ArcGIS Pro, open Vineyard.aprx from your GTKAGPro\Vineyard folder.

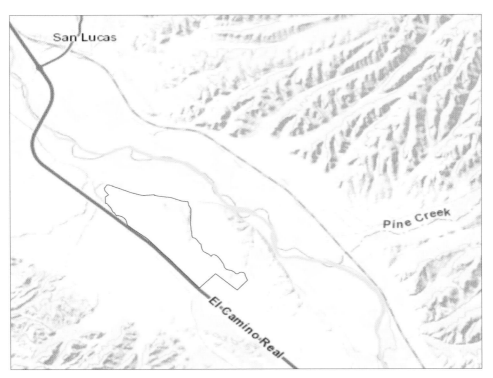

You see a basemap and a single line feature that demarcates the boundary of an approximately 1,100-acre property dedicated to viticulture.

2. Turn on the ned_1 and ned_2 layers.

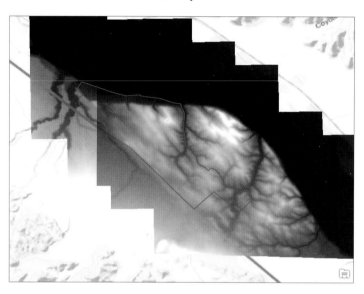

You will use the Extract by Mask tool to reduce the extent of the elevation layers to match the extent of the property boundary. However, to use the property boundary layer as a mask, it must be a polygon. Because it is currently a line feature (technically, three line features—you can tell by looking at the attribute table), you must first convert it to a polygon. For this task, you will use the Feature to Polygon geoprocessing tool.

3. In the Geoprocessing pane, search for and open the Feature to Polygon tool.

4. For Input Features, click the property_boundary line feature class.

    Notice that another input box appears—the tool provides the option to convert several feature classes at once, if needed.

5. Click inside the Output Feature Class box to see the default output location—the project geodatabase. Maintain this path but change the Output Feature Class name to **property_boundary_poly**.

6. Leave the optional parameters blank and run the tool.
    The new polygon layer is added to the map.

7. Remove the line layer, and symbolize the new polygon layer however you want.

Clip raster layers

Now that the property boundary is a polygon, you can move forward with clipping the elevation rasters.

1. In the Geoprocessing pane, search for and open the Extract by Mask tool.

2. For Input Raster, click ned_1. For Input Raster or Feature Mask Data, click property_boundary_poly.

3. Maintain the Output Raster name.

4. Run the tool. When the tool is finished running, turn off property_boundary_poly and ned_1.

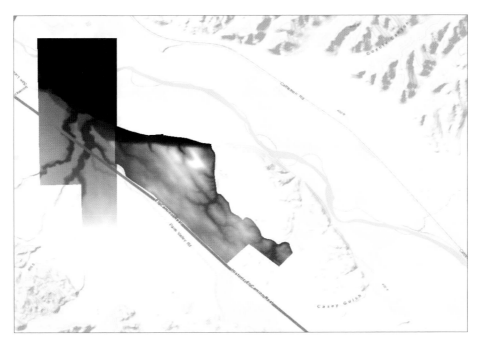

You have created a new raster whose shape matches the property boundary. Now you will perform the same operation on the second elevation raster.

*Another way to reduce the analysis extent is to use the Clip Raster tool in the
Data Management toolset. However, the tool clips only a rectangle, not a
polygon mask.*

5. Rerun Extract by Mask using ned_2 as the input raster and keeping property_bound-
ary_poly as the feature mask data. Rename the output raster **Extract_ned_2**.

6. Turn off ned_2 and zoom in slightly.

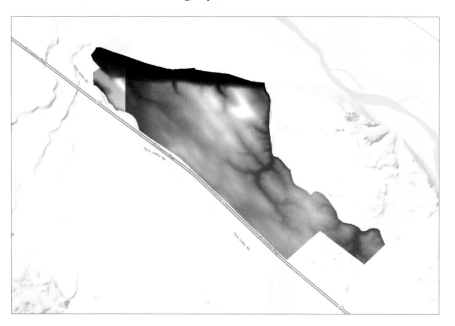

Merge rasters

Next, you will merge Extract_ned_1 and Extract_ned_2 into a new raster dataset so that you can
run analysis operations once, not twice.

1. Find and open the Mosaic to New Raster tool.

2. In the tool, set the following parameters:
- For Input Rasters, click Extract_ned_1 and click Extract_ned_2.
- For Output Location, browse to Vineyard.gdb, select it, and click OK.
- For Raster Dataset Name with Extension, type **ned_property**.
- For Spatial Reference for Raster, click property_boundary_poly.
- For Number of Bands, type **1**.

Why choose property_boundary_poly for the spatial reference? If you check the layer properties for the National Elevation Dataset (ned) rasters, you will see they are in a geographic coordinate system, but if you check the layer properties of the property boundary, you will see it has both a geographic and a projected coordinate system.

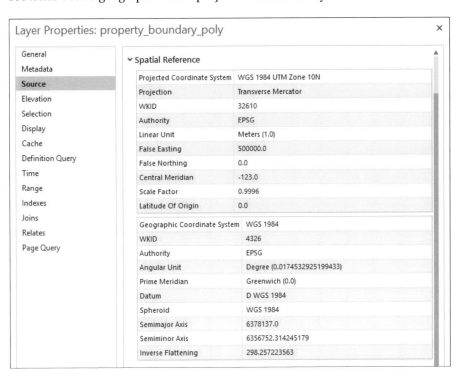

Before performing analysis, it is good practice to project your data to a local coordinate system. Recall that projections are discussed in chapter 4. With the Mosaic to New Raster tool, you are combining two rasters into a new raster dataset, and you are projecting the output, as well.

Learn more about the optional parameters by pointing to the information icons to read the ToolTips.

3. Run the tool. When processing is complete, remove all elevation layers except ned_property.

Notice that this new layer is a combination of the two inputs.

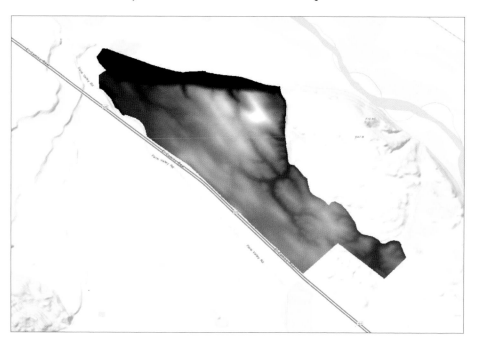

Your preprocessing is complete. One elevation raster is clipped to the extent of the study area and projected appropriately.

4. Save the project.

You have finished this exercise. The next exercise builds on the Vineyard project, so if you are continuing to exercise 8b, you can leave it open.

GIS IN THE WORLD: GIS AND PRECISION AGRICULTURE

Scheid Vineyards not only uses GIS for suitability analysis but also incorporates GIS and GPS tracking technology into nearly every aspect of its operation. From tracking and managing worker productivity and equipment locations to detecting and mapping harmful pests such as insects and mold, its comprehensive geospatial technology program informs decision-making from the ground up.

Figure 8.4. Scheid Vineyards. Courtesy of Heidi Scheid.

Read more about Scheid Vineyards' innovative approach to viticulture in *ArcNews*, "World-Class Vineyard Uses GIS to Fine-Tune All Its Operations": www.esri.com/about/newsroom/arcnews/world-class-vineyard-uses-gis-to-fine-tune-all-its-operations.

Exercise 8b: Derive new surfaces

Estimated time to complete: 30 minutes

Now you are ready to put some precision agriculture techniques to the test.

Exercise workflow

- Create an aspect surface to determine which areas of the property are south facing (the preferred aspect for a vineyard).
- Create a slope surface to determine which areas of the property have slope measurements of 14 percent or less (required for machine harvesting).
- Create three hillshade surfaces using almanac values for sun altitude and azimuth to determine which areas are in sun with minimal shadows at three different times of the afternoon in mid-September (longer sun ripening just before harvest is ideal).
- Overlay polygon features that show which areas are already planted to reveal the areas that are available for a new vineyard. Assess the suitability of these areas based on your aspect, slope, and hillshade layers.

Derive an aspect surface

1. Open the Vineyard project that you started in exercise 8a.

   If you completed exercise 8a, your project has a map named Suitability. The original two elevation rasters have been reduced to the extent of the property boundary (the project study area), merged into a single raster, and projected into a local coordinate system. The project includes a folder connection to the project folder, and the geoprocessing history reflects the work you have done thus far.

   If you did not complete exercise 8a or are not sure if you did it correctly, you can use Vineyard_ B.aprx, found in the Vineyard\Results\Vineyard_B folder.

   First, you will analyze the *aspect* of the topography (ground surface relief), to determine which direction each part of the ground primarily faces—north, south, east, west, or in between.

2. In the Geoprocessing pane, find and open the Aspect tool (there are two—open the one from the Spatial Analyst toolbox).

The Aspect tool in the 3D Analyst toolbox is the same tool. The Spatial Analyst and 3D toolboxes have quite a bit of overlap.

3. For Input Raster, click ned_property. For Output Raster, maintain the default location, but edit the name to **Aspect_ned**, maintain the remaining default parameters, and run the tool.

4. When processing is complete, examine the results.

    Notice the areas that face south, southeast, or southwest.

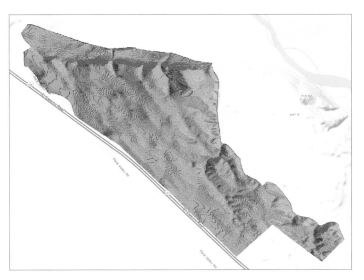

    Next, you will derive the slope layer.

Derive a slope surface

1. Find and open the Slope tool from the Spatial Analyst toolbox.

2. In the tool, set the following parameters:
    - For Input Raster, click ned_property.
    - For Output Raster, maintain the default location, but edit the name to **Slope_ned**.
    - Change the Output Measurement to Percent Rise.
    - Maintain the Method and Z Factor as is.

3. Run the tool.

4. When processing is complete, examine the results.

   You are interested in seeing only whether the slope is less than or greater than 14 percent, so you will symbolize the layer for only these two categories.

5. With Slope_ned selected in the Contents pane, open the Symbology pane.

6. In the Symbology pane, change Method to Equal Interval and the number of classes to **2**.

7. After you change the classes to 2, change Method to Manual Interval.

8. For Color Scheme, click the down arrow, check the box for Show Names, and click the Slope color scheme.

9. At the bottom of the pane, on the Classes tab, click the first Upper Value box once to select it, and click again to make it editable. Replace the existing value with **14**, and press Enter.

   You can see that although much of the property meets the criterion of having a slope that is less than or equal to 14 percent, a fair amount of it still has a greater percentage rise, which is not ideal for a new machine-harvestable vineyard.

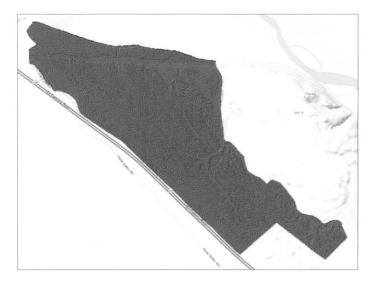

Derive a hillshade surface

A *hillshade* is a surface layer that depicts shadows to model the effect of an illumination source (usually the sun) over the terrain of the land. To create a hillshade, you must enter an azimuth and altitude value. *Azimuth* is the direction of the sun, expressed in positive degrees from 0 to 360, measured clockwise from north. The default azimuth is 315 degrees. *Altitude* is the angle of the sun above the horizon. The altitude is expressed in positive degrees, with 0 degrees at the horizon and 90 degrees directly overhead. In other words, relative to any given surface, azimuth tells you where in the sky the sun is, and altitude tells you how high. These values change depending on the time of day and the time of year.

Your criteria indicate that you want a lot of sun, especially during the critical fruit-ripening stage just before harvest. Harvest is usually sometime in October, depending on various factors. To be precise, you will look up almanac values for this area on September 15, at 2:00 p.m., 3:00 p.m., and 4:30 p.m., to show which areas have the longest sun exposure at this time.

1. Find and open the Hillshade tool in the Spatial Analyst toolbox.

2. For Input Raster, click ned_property. For Output Raster, maintain the default location, but edit the name to **Hillshade_ned_1**.

    You will use the azimuth and altitude values from the US Naval Observatory Astronomical Applications Department.

3. Enter the following values in the Hillshade tool: azimuth of **206.2** and altitude of **53.9**. Maintain the remaining default settings and run the tool.

The output is a shaded relief layer that shows which areas are illuminated and which are in shadow at 2:00 p.m. in mid-September. A raster cell value of 0 represents complete shadow, with no illumination from the sun. A raster cell value of 255 represents full illumination.

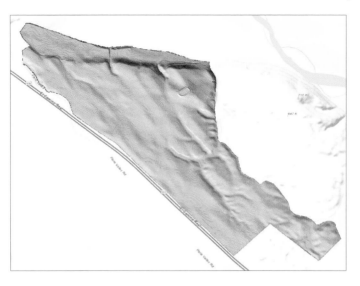

Is any part of the property in complete shadow at this time?

4. Repeat the hillshade process two additional times, using the following values for 3:00 p.m. (15:00) and 4:30 p.m. (16:30): 3:00 p.m. azimuth of **226.8** and altitude of **46.6**, and 4:30 p.m. azimuth of **248.1** and altitude of **31.3**. Name the output rasters **Hillshade_ned_2** and **Hillshade_ned_3**, respectively.

The illuminated areas decrease as the sun descends in the sky.

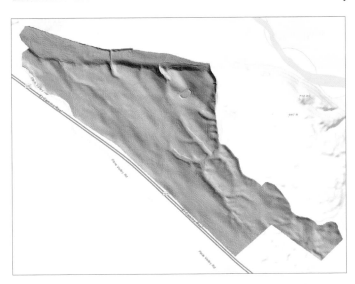

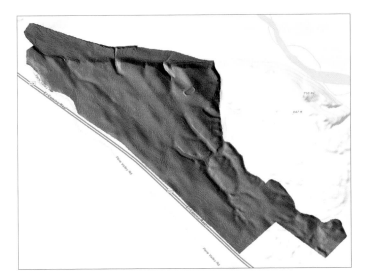

Hillshade rasters are natural candidates for 3D display.

5. Turn off all other layers except Hillshade_ned_3 and the basemap.

6. On the View tab, click Convert, and choose to convert to a local scene.
   This may take some time to draw and is not necessary to complete the exercise.

7. When the Suitability_3D map is created, use the Explore tool to rotate and tilt the map so you can see the relief.

> **REMIND ME HOW**
>
> Press and hold the scroll wheel on the mouse to rotate and tilt a 3D map.

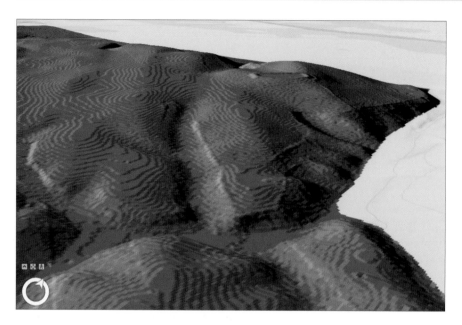

You will complete the chapter in the 2D map, but feel free to examine the 3D map at any time.

8. Close the Suitability_3D scene when you have finished.

Visually compare analysis outputs

1. In the Suitability map (the 2D map), add vineyard_blocks from Vineyard.gdb. Symbolize it in brown with a black outline.

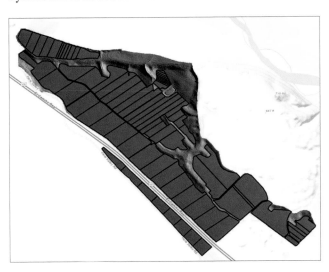

This layer represents already planted vineyard blocks. Next, you would typically hand-digitize a layer of potential planting sites using ArcGIS Pro editing tools and your knowledge of the property, but to save time, we have provided the layer for you.

2. Add planting_sites from Vineyard.gdb, and symbolize it using a black 2-point outline and no color fill for easy visibility.

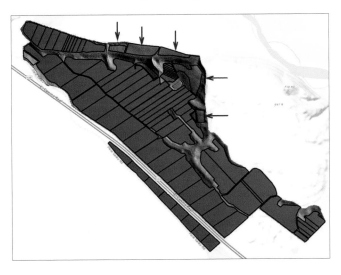

1 2 3 4 5 6 7 8 9

These unplanted areas of the property are available for future plantings. First, you will assess the slope criteria because machine harvesting (rather than picking by hand) is a requirement in this scenario.

3. Turn off all hillshade layers, turn on Slope_ned, and look for future planting sites that have mostly low-slope land (a lot of green shading).

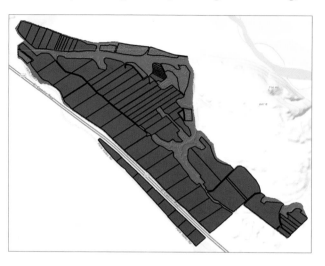

How many planting sites contain mostly low-slope (less than 14 percent) topology?

4. Turn off Slope_ned to reveal the Aspect_ned layer.

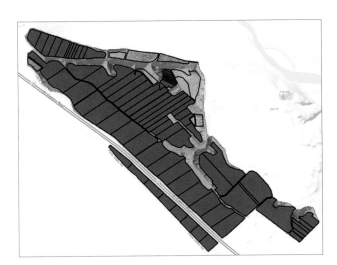

How many potential planting sites include at least some land that faces
south, southeast, or southwest? (Zoom in and pan as needed.)

5. Turn on hillshade_ned_1.

Are any of the potential planting sites in shadow
at 2:00 p.m. in mid-September?

6. Turn on hillshade_ned_2. Notice that the shadows intensify over most of the planting
sites.

7. Turn on hillshade_ned_3. Notice that the shadows intensify further in most areas.

Can you identify a planting site that meets the slope and
aspect criteria and has decent sun exposure at 4:30 p.m.,
thus revealing the best site to plant the new vineyard?

Rarely will a site fit all your suitability criteria perfectly. In exercise 8c, you will combine your criteria in a weighted model for a more precise analysis.

8. Save the project.

If you are continuing to exercise 8c, keep the project open. Otherwise, you can exit ArcGIS Pro and come back to it later.

Exercise 8c: Create a weighted suitability model

Estimated time to complete: 20 minutes

You will find the most suitable planting site by combining the values of three layers (aspect, slope, and the third hillshade that shows sun illumination in the latest part of the afternoon). Before the layers are combined, they must have a common range. Combining them as they are now would produce nonsensical results. Consider: How does a slope of 9 compare with a hillshade value of 190? Instead, you will *reclassify* the values so that you can add them together. (That means you will replace raster cell values with new cell values so that the rasters can be logically combined.) Each layer will be weighted—that is, as a percentage of the whole. In this way, you can prioritize your criteria. For example, if having the correct slope is most important, you will give slope the highest weight.

Exercise workflow

- Create a new geoprocessing model. Add three Reclassify processes, in which you will reclassify three criteria surfaces: Aspect_ned, Slope_ned, and Hillshade_ned_3. Replace the original cell values with either 1 (for desirable values) or 0 (for undesirable values).
- Add a Raster Calculator process to the model. Add together the reclassified values of the three inputs but multiply each input by a percentage of 100. In this way, your end results will be weighted according to criteria priority.

Reclassify criteria rasters

1. In ArcGIS Pro, continue working with Vineyard.aprx.

If you completed the previous exercises in this chapter, your Suitability map contains the topographic basemap and nine layers, including three feature classes (planting_sites, vineyard_blocks, and property_boundary_poly, which is turned off), plus three hillshade surfaces, a slope surface, an aspect surface, and the ned_property elevation surface from which the other surfaces were derived.

If you did not complete exercises 8a and 8b or are not sure if you did them correctly, you can use Vineyard_ C.aprx, found in the Vineyard\Results\ Vineyard_C folder.

2. On the Analysis tab, click the ModelBuilder button to create a new geoprocessing model in the project toolbox.

Before you can combine the criteria into a single output, the surfaces must be reclassified so that their cell values make sense in a calculation. In this case, you will change the desirable cell values to 1 and the undesirable values to 0.

3. Search for the Reclassify tool (Spatial Analyst tools) in the Geoprocessing pane, but do not open it. Drag the tool from the list of search results into the model three times.

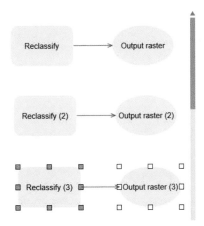

The ModelBuilder tab has a Tools button, which allows you to open the Geoprocessing pane from there.

First, you will reclassify the aspect surface.

4. Switch back to the Suitability map and examine the Aspect_ned legend.

The desirable values are Southeast (112.5–157.5), South (157.5–202.5), and Southwest (202.5–247.5). You will also consider flat surfaces (–1) as desirable.

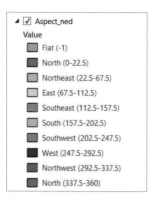

5. Switch back to the Model view. In the model, double-click the first Reclassify tool to open it.

   - For Input raster, click Aspect_ned.
   - Maintain the Reclass Field of VALUE.
   - Give the desirable traits (Flat, South, Southeast, and Southwest) a new value of **1** and the others a new value of **0**.

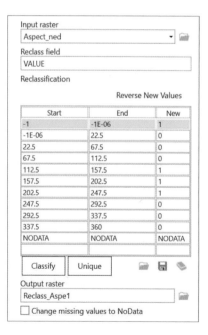

Be careful that your screen matches the figure correctly, or your final output may be incorrect.

6. Maintain the default output and click OK.

The tool is now ready to run in ModelBuilder.

Resize or move the model elements as you prefer. You can also click the Auto Layout button on the ModelBuilder tab in the View group to arrange model elements automatically.

7. Open the second Reclassify tool in the model.
- For Input Raster, click Slope_ned.
- Maintain the other parameters, but for the 14–143.770981 category, change the New value to **0**.

8. Click OK to make the tool ready in the model.

The second tool is ready to run.

9. Open the third Reclassify tool.

Before you set the parameter for Input Raster, you will set the Reclassification values.

The value table currently has no values. The reason is because the hillshade layers are symbolized using a stretched symbology scheme, so there is not a category for each range of values. You know that some shade is unavoidable at 4:30 in the afternoon, yet you want to rule out areas that have little to no sunlight. Knowing that 0 means no illumination and 251 is the highest level of illumination at this time of day (look at the legend), you decide to draw the "unacceptable" cutoff value at 90.

10. In the first Reclassification cell under Start, type **0**. Under End, type **90**, and under New, type **0**.

11. Press Enter to start a new row. In the next row, under Start and End, type values of **91** and **251**, respectively. Under New, type **1**.

12. For Input Raster, click Hillshade_ned_3.

13. Make sure that your screen matches the graphic, and then click OK.

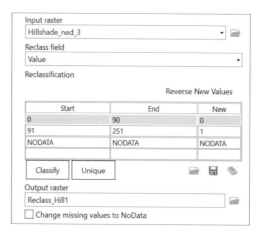

All the Reclassify tools are ready to run.

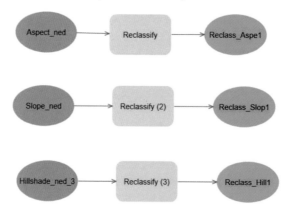

Combine criteria rasters

1. Search for the Raster Calculator tool in the Geoprocessing pane (Spatial Analyst toolset) and add it to the model.

You decide to weight aspect, slope, and hillshade to 20 percent, 50 percent, and 30 percent, respectively.

2. Open the Raster Calculator tool. Double-click Reclass_Aspe1 to add it to the expression box.

In the expression box, it looks like "`%Reclass_Aspe1%`". This output raster is from the operation that will reclassify the aspect layer.

3. Place your pointer after the final quotation mark and double-click the multiplication operator (*) to add it to the equation. Enter a space, and type **.20**.

4. Double-click the addition operator (+).
So far, the expression reads as follows: "%Reclass_Aspe1%" * .20 +

5. Double-click Reclass_Slop1 to add it to the expression. Double-click the multiplication operator, type a space, and type **.50**.

6. Double-click the addition operator.

7. Double-click Reclass_Hill1. Double-click the multiplication operator, type a space, and type **.30**.

8. Set the Output Raster name to **rastercalc**. Verify your algebra expression.

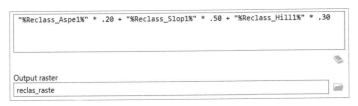

```
"%Reclass_Aspe1%" * .20 + "%Reclass_Slop1%" * .50 + "%Reclass_Hill1%" * .30
```

Output raster
reclas_raste

9. Click OK.

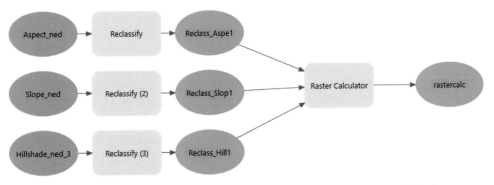

10. Save and run the model. (If you want, you can also validate your model before running.)

11. When processing is complete, close the progress window.

12. In the Catalog pane, notice that a new raster named rastercalc is in your project geodatabase. (You may need to refresh the geodatabase first to see it.)

13. Click the Suitability map tab.

14. From the Catalog pane, drag rastercalc to the map below the planting_sites layer. Click Yes on the Build Pyramids for Rastercalc notification.

15. Turn off vineyard_blocks.

16. Symbolize rastercalc using classified symbols, Natural Breaks method, 3 classes.

Areas of darker red indicate higher values from the raster calculation and the most desirable characteristics. The three areas at the top of the image are the top three candidates.

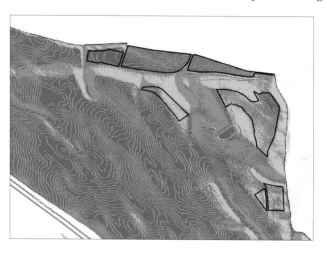

17. Save the project.

The GIS analysis is an important first step, but it is just one piece of the equation. You know that these sites are geographically suitable for planting, but now the owners must consider several other factors before investing in planting. For instance:
- What grape varietal makes the most sense from a business perspective? (Syrah, pinot noir, and chardonnay are common varietals in this region.)
- What is the average tonnage yield per acre? (Because of your scrupulous suitability analysis, the yields are likely to be at least average, likely better.)
- Are these grapes going to be sold to other winemakers or used for the owner's label?
- If the former, what is the average price per ton that they can expect to reap?

- If the latter, how many cases is the owner's winery able to get per ton of grapes, and what is the profit margin?
- How much will the additional vineyard blocks add to the overhead costs of fieldworkers, vineyard managers, fertilizer, water, pest abatement, and so on?
- Will the financial gain in good years be enough to offset losses from the inevitable challenges of drought, pests, or early frost?
- What type of soil exists at each site, and how does that type affect the decision? (Soils are important in terms of rootstock selection, which down the line will dictate your irrigation practices.)

ON YOUR OWN

Add Soils_clip from Vineyard.gdb. Symbolize it using unique values based on the Name value field. Place the planting_sites layer at the top of the Contents pane.

What soil types are represented in your top-choice planting sites?

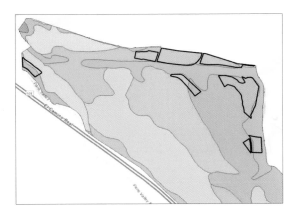

The work you have done thus far, using ArcGIS Pro and Spatial Analyst, provides an excellent starting point in the decision-making process with the vineyard management team.

ON YOUR OWN

Switch out the topological basemap for an imagery basemap, and turn off all layers except planting_sites.

Have any of the potential sites been planted?

You are finished with this chapter. If you are not continuing to chapter 9, you can exit ArcGIS Pro.

Summary

This chapter gave you an introduction to precision agriculture using GIS. Using raster elevation data and the Spatial Analyst extension, you created aspect, slope, and hillshade layers. Then you learned how to create a symbol model that combined the three criteria to determine the best planting site. Remember that before combining rasters using the Raster Calculator tool, you will likely have to reclassify them to give them a common value scheme. You can see that for some GIS purposes, Spatial Analyst is an indispensable tool.

Glossary terms

cell
discrete data
continuous data
map algebra
NoData
mask
aspect
hillshade
azimuth
altitude
reclassify

CHAPTER 9
Presenting your project

Exercise objectives

9a: Apply detailed symbology
- Create a definition query.
- Fine-tune the symbology.
- Apply symbol layer drawing.

9b: Label features
- Label features using Maplex Label Engine.

9c: Create a page layout
- Add a layout to the project.
- Insert map frames.

- Insert a legend element.
- Insert a scale bar.
- Insert a north arrow.
- Insert dynamic text, a title, and rectangles.

9d: Share your project
- Export a layout to a PDF file.
- Save a layout file.
- Package and share a project template.

Maps convey your ideas about locations to the minds of map readers. It is a fine balance of communicating the appropriate amount of information by displaying data and graphic elements through a suitable medium, such as a paper map or mobile device.

Every *layout* has a collection of organized elements. In a layout, you can arrange common elements, including the map, map labels, a title, legend, scale bar, north arrow, captions, and additional graphics. Creating a page layout requires answering some key questions:

- What is the purpose and intended use of the map?
- Who is the target audience, and what is their familiarity with the area of interest? What kinds of labels will provide a sense of familiarity?
- What is the most appropriate map scale to convey the map information? How detailed or general will the map data be displayed?
- What type of map is it—quantitative or qualitative? What kind of symbology will most effectively communicate the data?
- What are the dimensions of the map (printed or digital)? What display orientation will best fit, and how will the map elements be organized?

A good map not only provides facts but also persuades opinion. You will want to think about the context of the data and frame the results to match the geographic area of the phenomena. Also think about graphic hierarchy and how map features are overlaid on top of each other. A well-composed layout promotes a visual hierarchy that map users can explore naturally. Remember that print maps are just one way to share your work and that ArcGIS Pro provides other options, including sharing as a PDF, a web map, or a web scene in the case of 3D.

In this chapter, you will compose a layout that consists of several maps of broadband internet availability for two counties in Utah.

Scenario: The agency you work for aggregates much of the broadband accessibility data for the state of Utah. As part of the agency's ongoing activities, it will determine broadband internet availability and work with stakeholders and service providers to identify accessibility, support broadband planning and policy, and study broadband trends.

Much of the broadband internet availability data has been collected through broadband service provider submissions and publicly available sources. As part of a large mapping project, several areas must be mapped. You will use your knowledge of cartographic principles and GIS skills to complete maps that show the maximum advertised broadband speeds for a county in Utah. In addition, you will prepare some data before symbolizing and labeling it. Once you have identified which areas to map, you will create several layouts to complete the mapping project, which can then be exported and shared online.

GIS IN THE WORLD: TELECOMMUNICATIONS

TK Telekom is a dynamically developing telecommunications operator that uses Esri and ArcGIS solutions to efficiently manage its network of nearly 30,000 kilometers of lines. The company provides internet services, telephone services, data transmission, and line lease for telecommunications operators, public administrators, and business customers. GIS operators can see the inventory system quickly to identify potential failure locations and enable implementation of precise remedies against a background of digital vector and raster maps. Read more about the project in *ArcNews*, "Facilitating and Improving Telecom Network Management": www.esri.com/esri-news/arcnews/winter1213articles/facilitating-and-improving-telecom-network-management.

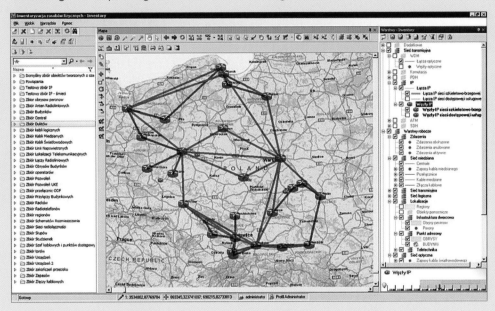

Figure 9.1. The SunVizion Network Inventory, based on ArcGIS Server and Microsoft SQL Server platforms, visualizes this IP network. Courtesy of Suntech S.A.

DATA

In the GTKAGPro\Broadband\UtahBroadband.gdb:

- Wireline_Broadband–a polygon feature class of wired internet coverage *(Source: Utah Broadband Project; Utah Automated Geographic Reference Center; Governor's Office of Economic Development)*
- Wireless_Broadband–a polygon feature class of wireless internet coverage *(Source: Utah Broadband Project; Utah Automated Geographic Reference Center; Governor's Office of Economic Development)*
- Counties–Utah county boundaries *(Utah AGRC, BLM)*
- Community_Anchor_Institutes–a point feature class of the community sites *(Source: Utah Broadband Project; Utah Automated Geographic Reference Center; Governor's Office of Economic Development)*

Exercise 9a: Apply detailed symbology

Estimated time to complete: 20 minutes

In this exercise, you will apply detailed symbology to the data to visualize maximum advertised broadband speeds—both wired and wireless broadband. To view symbology and features, it is best to maximize your ArcGIS Pro window, if possible.

Exercise workflow

- Create a definition query to show only fixed wireless technology.
- Change symbology to symbolize only certain broadband download speeds.
- Apply symbol layer drawing to control the drawing order of wired broadband symbology.

Create a definition query

1. Open the Broadband.aprx project from the GTKAGPro\Broadband folder.

   The project has two maps, Max Advertised Broadband Speed and Community Anchor Institutes, which you will use later. In the active map, the Wireline_Broadband layer represents all census tracts that have wired broadband technologies, and the Wireless_Broadband layer represents areas that have wireless technologies. You are required to show only fixed wireless technology and visualize it. To show only fixed wireless, you will create a definition query to select a subset of all the wireless technologies.

2. In the Symbology pane, observe how the Wireless_Broadband layer is symbolized.

The Wireless_Broadband layer is symbolized using Unique Values and a random color scheme. This layer has its values stored as numbers in the geodatabase and wireless broadband technology label names associated with them. You are interested in Terrestrial Fixed Wireless technologies that have corresponding values of 70 and 71, Unlicensed and Licensed, respectively.

> **ON YOUR OWN**
>
> Open the Wireless_Broadband attributes, and explore the geodatabase fields, domains, and subtypes in Fields View to get an idea of how the data is organized.

3. With the Wireless_Broadband layer selected in the Contents pane, go to the Data tab on the ribbon and click the Definition Query launcher.

Creating a definition query expression to select a subset of features for the layer to display is slightly different from using the geoprocessing tool Select Layer by Attribute. Fundamentally, creating the expression is the same, but the latter results in selected features. In this workflow, only the features that satisfy the expression are displayed on the map although they are not selected.

4. On the Definition Query tab, click New Definition Query, and create the following expression: **Where Technology_of_Transmission is Equal to 70 – Terrestrial Fixed Wireless – Unlicensed**.

5. Click Add Clause.

6. Add a second clause using an OR operator that queries the value **71 – Terrestrial Fixed Wireless – Licensed**.

Using the AND operator will not work here, because in this dataset there are two distinct fixed wireless values in the Wireless_Broadband layer. When the OR operator is used, you can select from all points—that is, Terrestrial Fixed Wireless – Unlicensed or Terrestrial Fixed Wireless – Licensed.

7. To view the SQL statement in full, turn on the SQL toggle button in the query dialog box.

TRANSTECH is the field name of the Technology of Transmission field.

8. Click the green check mark to verify the SQL expression. (If the SQL expression was not verified successfully, retrace your steps, and try again.) Click Apply and click OK to close the dialog box.

A successful definition query results in only fixed wireless technology being displayed on the map.

9. Turn off the Wireline_Broadband layer.

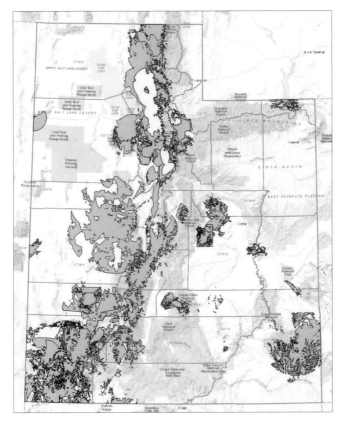

With the Wireline_Broadband layer turned off, you can see the Wireless_Broadband layer more clearly.

How many areas are fixed wireless technology?

Fine-tune the symbology

After the definition query isolates only fixed wireless technology areas, you will apply more detailed symbology to the wireless broadband layer to show the maximum advertised download speed available. You will symbolize only broadband download speeds that are faster than 768 kilobits per second (Kbps).

1. With the Wireless_Broadband layer selected in the Contents pane, go to the Symbology pane and make sure Unique Values is chosen.

2. For Field1, click Maximum_Advertised_Downstream_Speed. (If necessary, adjust the Label column width to make the labels legible.)

3. Right-click the Value column heading and click Sort Ascending.

Eleven values are shown. This will provide a rough ranking of the values. Several values do not meet the requirement of symbolizing speeds of only 768 Kbps or faster. You will remove them and effectively hide the others.

4. Right-click and remove value 1 (Less than or equal to 200 Kbps).

5. Remove value 2 (Greater than 200 kbps and less than 768 Kbps).
Next, organize the categories into a more logical order.

6. Sort the Value column in descending order.

The values 10 and 11 are at the bottom of the list. Because they are both categorized as faster (gigabits per second [Gbps]) download speeds, you will move them to the top of the list.

7. Select both values 10 and 11 (press Ctrl while clicking each row), right-click and click Reorder > Top.

Nine values are a lot to symbolize. Next, you will group the three highest values together to reduce the number of total values to just seven values.

Remember that panes can be either docked or floating. You can adjust their width or height to make pane elements visible and easier to work with.

8. Select values 9, 10, and 11, right-click and click Group Values. Rename the group **Greater than 50 mbps**. Rename the remaining values, using the following ranges in the text descriptions:
- Value 8: **25 to 50 mbps**
- Value 7: **10 to 25 mbps**
- Value 6: **6 to 10 mbps**
- Value 5: **3 to 6 mbps**
- Value 4: **1.5 to 3 mbps**
- Value 3: **768 kbps to 1.5 mbps**

Symbol	Value	Label
⌄ **Maximum_Advertised_Downstream_Speed**		7 symbol classes •••
☐ ▾	9; 10; 11	Greater than 50 mbps
☐ ▾	8	25 to 50 mbps
☐ ▾	7	10 to 25 mbps
☐ ▾	6	6 to 10 mbps
☐ ▾	5	3 to 6 mbps
☐ ▾	4	1.5 to 3 mbps
☐ ▾	3	768 kbps to 1.5 mbps

The seven values have been prepared and are now ready for a different color scheme.

9. From the Color Scheme list, click the Show All and Show Names check boxes. Scroll down and click Red-Yellow-Green (7 Classes).

10. On the Classes tab, click More and click Format All Symbols.

11. Go to Properties, change the Outline color to No Color, and click Apply.

12. Click the back button to return to the Wireless_Broadband Symbology pane. Click More and click Show All Other Values to disable it.

13. On the ArcGIS Pro ribbon, click the Feature Layer contextual tab, locate the Effects group, and adjust the layer transparency to **40%**. (Alternatively, type **40**, and press Enter.)

14. Turn on the Wireline_Broadband layer.

The symbology for the Wireless_Broadband layer is complete.

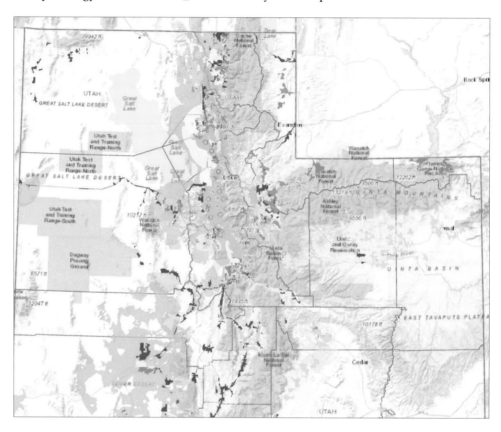

Apply symbol layer drawing

Next, you will look at the symbology on the map and compare it with the order of the color ramp in the Contents pane. It may be difficult to see how overlapping features are being represented, but when you inspect it closely, the higher speeds are drawn in a lower order.

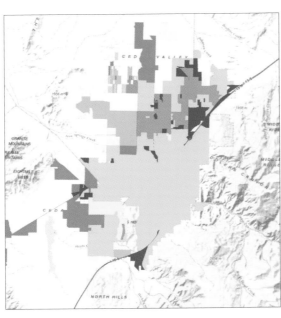

Figure 9.2. In this map, no symbol layer drawing is used.

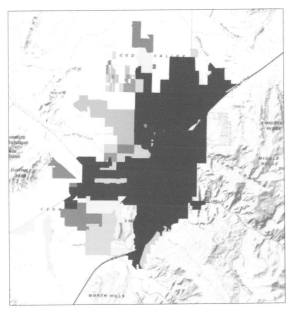

Figure 9.3. In this map, symbol layer drawing is enabled. The symbols are drawn according to the order specified in the Symbology pane.

You will use symbol layer drawing to control the drawing order of feature symbology—essentially overriding the default drawing sequence. This override is particularly helpful in this situation in which overlapping features exist.

1. With the Wireline_Broadband layer selected in the Contents pane, go to the Symbology pane and click the Symbol Layer Drawing tab at the top. (Point to each of the five symbols to locate the correct tab.)

2. Turn on the Enable Symbol Layer Drawing toggle button.

3. Expand any group to see its contents.

   You now have the correct drawing order based on the order that was used in the Symbology pane, so no reordering is needed.

4. Turn off the Wireline_Broadband layer.

   The current drawing order for the Wireless_Broadband layer does not follow the desired hierarchy.

5. Select the Wireless_Broadband layer and repeat steps 1 to 3 to enable symbol layer drawing.

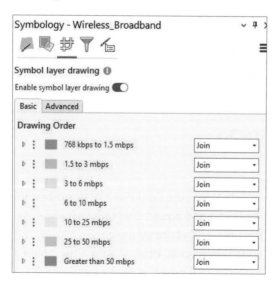

Enabling symbol layer drawing has overridden the default drawing order. Now you can manually set the desired order.

6. Drag the color swatches to set the drawing order to draw the slowest speeds first (lower) and have them overlapped by the fastest speeds (higher).

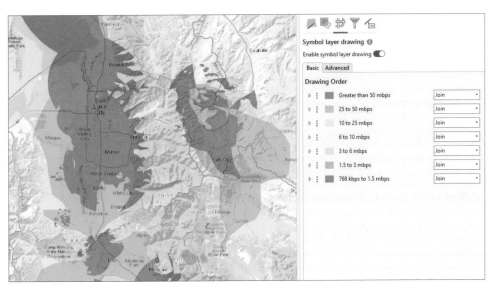

7. Turn on the Wireline_Broadband layer.

   This map now shows coverage by fixed broadband technologies (DSL, cable, fiber, and fixed wireless) based on maximum advertised download speeds marketed to all consumer levels, including commercial businesses. Next, you will label features in another map before you use it in a page layout.

Exercise 9b: Label features

Estimated time to complete: 15 minutes

Your mapping project also requires features to be labeled on the Community Anchor Institutes map. *Labels* are based on one or more feature attributes and placed near or on a feature. As you may have seen in earlier chapters, ArcGIS Pro places labels for all features in a layer with a single click based on predetermined labeling rules. This process, known as dynamic labeling, is quick but has its drawbacks. Label positions can change depending on map scale, addition or removal of features, or the attribute used. Dynamic labels are still useful for most mapping projects. However, keep in mind that static layouts for print require extra attention so that labels are exactly where you want them.

MAPLEX LABEL ENGINE VERSUS STANDARD LABEL ENGINE

Maplex™ Label Engine is the default label engine, because it provides the best positioning and has finer control over labels, such as enhanced position strategy, conflict resolution, weighting, and priority. Standard Label Engine is used to label as many features as it can without overlap and for cases in which the utmost speed is needed. To learn more about the differences between the two label engines, go to ArcGIS Pro Help > Maps and Scenes > Author Maps and Scenes > Text > Text on a Map.

The map has several layers, including a point layer that represents community libraries. It is symbolized based on whether there is public Wi-Fi available at each location. However, some libraries are symbolized as unknown. You will create labels so that your department can get a visual perspective of which libraries still require information to be collected so that the database can be updated accordingly.

Exercise workflow

- Use Maplex label settings to label libraries that have an unknown public Wi-Fi status.

Label features using Maplex Label Engine

When you use Maplex Label Engine, you have access to a new set of label placement properties that allow you to control how labels are oriented, formatted, and placed in feature-dense areas and how conflicts between labels can be resolved. In addition to the standard feature types, Maplex Label Engine provides label placement options for features such as streets, contours, rivers, boundaries, and land parcels. You will use some of these capabilities to label libraries.

1. Ensuring that Community Anchor Institutes is the active map, on the Map tab, go to the Utah County bookmark.

You will concentrate on labeling libraries in Utah County. Points that are densely located are more difficult to label, and you must take into consideration such factors as map scale, label font size, color, and placement.

2. In the Contents pane, select the Libraries layer.

3. On the Labeling tab, in the Map group, click More to ensure Use Maplex Label Engine is checked. If it is not enabled, click Use Maplex Label Engine to enable it.

4. On the Labeling tab, in the Layer group, click the Label button. (Alternatively, right-click the Libraries layer and click Label.)

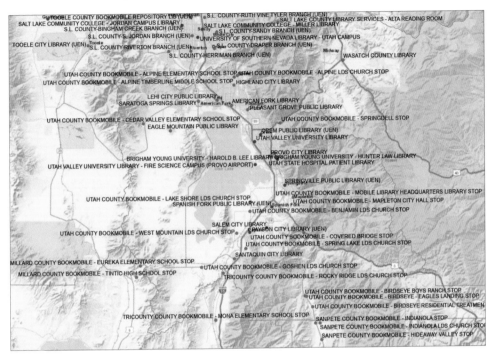

Labels are placed on the map for every feature in the layer. But labeling, in this case, is not helpful because there are too many labels. The initial label placement is based on default Maplex Label Engine rules using the Class 1 *label class*. Label classes are used to specify detailed aspects of how labels are positioned and symbolized. You will modify the label class so that it labels only those libraries that have an unknown public Wi-Fi status.

5. In the Label Class group on the ribbon, click SQL Query to open the Label Class pane.

6. On the SQL Query tab, build a query in which PublicWifi is equal to Unknown. Click Apply.

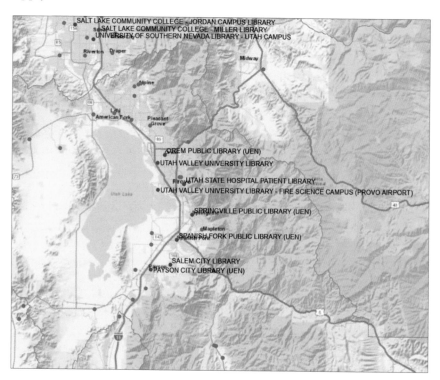

ON YOUR OWN

Optionally, use a label expression to change the labels from all uppercase letters to title case. A label expression is used to format labels using Python, VBScript, or JScript logic. Try applying the Python expression `[ANCHORNAME].title()` to the label class. The `.title()` text function formats the labels to title case (for example, `SALEM CITY LIBRARY` becomes Salem City Library).

The total number of labels has been reduced, and only labels for libraries that have an unknown public Wi-Fi status are shown. Next, you will adjust the label class symbol properties.

7. In the Label Class pane, click Symbol.

8. On the General tab, expand Appearance. Change the font to Calibri, the font style to Bold, the size to **8** pt, the color to Gray 60%, and click Apply.

9. Collapse Appearance, and expand Halo. For Halo Symbol, click White Fill 50% Transparency and click Apply.

The labels now look more legible on the map. However, they still look clustered closely together. You will adjust the label class position next.

10. In the Label Class pane, on the Position tab, expand Placement. Ensure that Best Position is selected in the label placement list. Change Measure Offset From to Exact Symbol Outline.

11. On the Fitting Strategy tab, expand Stack. Change Maximum Number of Lines to **2**.

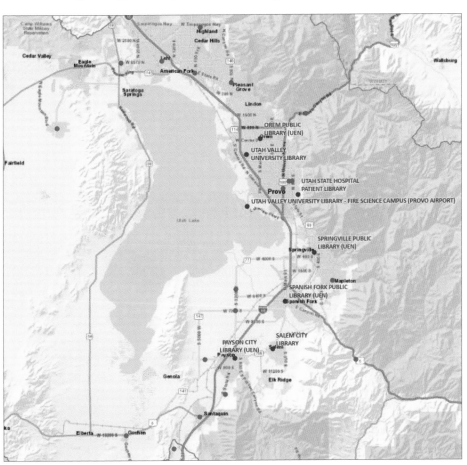

The labels now look properly placed without overcrowding. Remember that the entire map is labeled in the same way. If you pan or zoom to see the rest of the map, the labels will dynamically redraw to label all visible features based on this label class. To show labels at only specific map scales, you can specify a label visibility range.

12. On the Labeling tab, type **1:450,000** in the Out Beyond box, and press Enter.

When the map is zoomed out beyond 1:450,000, the labels are not displayed.

You will create another label class to label more features, this time for Utah counties. This label class will provide more visual information to your map users.

13. Enable labeling for the Counties layer.

14. In the Label Class pane, click Class and click the pane options menu. Rename the label class **County Boundary**.

15. Click Symbol and expand Appearance. Change the font to Century, the font size to **7** pt, and the color to Raw Umber (five rows down and four columns from the left). Click Apply.

16. Click the Position tab, click the Position button, and expand Placement. (Do not get confused with expanding Position under Symbol.) Click Boundary Placement.

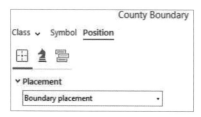

The county labels are switched from the default regular placement to boundary placement, which means that they follow the boundary lines close. This type of labeling allows readers to easily locate boundaries on a map.

17. On the Conflict resolution tab, expand Repeat. Set Minimum Interval to **4.0** inches (you need to change the unit type as well).

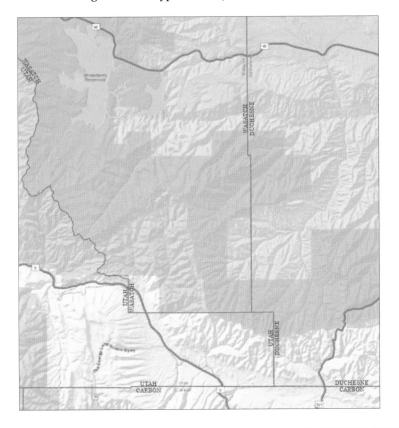

ON YOUR OWN

Create a new label class, and experiment with label placement, fitting strategies, and conflict resolution. The capability for fine-tuning labels is extensive and takes practice to get right. When you have finished, return to the County Boundary label class.

In exercise 9c, you will create a map layout that your readers can use to learn more about the broadband availability in different parts of Utah.

Exercise 9c: Create a page layout

Estimated time to complete: 45 minutes

A page layout includes map frames and other map elements, such as a scale, legend, north arrow, and more. Deciding the dimensions and orientation of a layout early on facilitates the design process. It influences your decisions on factors such as map frame sizes, map scale, text size, and how other map elements may fit.

ArcGIS Pro has rulers, guides, and a grid to help you arrange map elements on a page. You can also align, nudge (move incrementally), distribute (space evenly), order, rotate, and resize selected elements to place them where you want them.

For this exercise, you will insert map frames into the layout and add surrounding map elements that will help your readers.

Exercise workflow

- Add a new layout to the project based on an existing size.
- Insert two map frames and position them side by side.
- Insert a legend based on the symbology used for each map frame.
- Insert a scale bar for each map frame.
- Insert a north arrow.
- Add text and a title to the layout.

Add a layout to the project

1. On the Insert tab, in the Project group, click New Layout.

2. In the gallery (from the ANSI – Landscape group), click the Letter 8.5" x 11" page size.

   A landscape layout is added to the project. Like maps and data, layouts are stored in the project. You can access a layout at any time from the Project pane in the Layouts folder. A contextual tab named Layout appears on the ribbon and contains layout-specific tools. The other tabs also contain contextual layout-specific settings.

To change the page size or orientation after adding it, go to the Layout tab and use the controls in the Page Setup group. You can also click Properties in that group (lower-right corner) and click Page Size to quickly change to another standard page size and orientation or right-click the layout name in the Contents pane to launch the Layout Properties dialog box.

Next, you will add content to your layout—including map frames, text, map surrounds, and graphics—using the Insert tab.

Insert map frames

Map frames are containers for maps in your page layout. They can contain any map in your project, including 3D scenes. In layout templates, the map frames can be left empty and used to point to a map later.

Changing the map extent or scale inside a map frame does not change the actual map that may already be open in your project.

1. On the Insert tab, in the Map Frames group, click the Map Frame down arrow, and choose Default Extent from the Community Anchors Institutes gallery.

2. Use your pointer to draw a rectangle and place the map frame in the layout.
   A map frame that contains the Community Anchor Institutes map is added to the page.

3. With the map frame selected, click the Map Frame contextual tab. In the Size & Position group, change the width to **5 in** and height to **6 in**. (The Width and Height controls are on the far right.)

4. In the Size & Position group, click the upper-right anchor, and change X and Y to **10.6 in** and **7.5 in**, respectively.

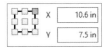

The Size & Position controls allow you to fine-tune placement on the page. Next, you will add a guide and the Max Advertised Broadband Speed map as a map frame.

5. Right-click the ruler at the top of the frame and click Add Multiple Guides.

6. In the Add Guides dialog box, ensure that Orientation is set to Vertical. For Position, type **0.375 in** and click OK.

A guide is placed at the 0.375-inch position of the layout. Guides are available to help you position frames and other elements in your layout. A useful feature is to snap layout elements to guides. When you move or resize any layout element near a guide, the element will snap to the guide.

7. Repeat the procedure to insert Default Extent from the Max Advertised Broadband Speed gallery as a map frame.

Do not worry about the look of the map. For now, you are only defining a placement within the layout.

8. Use the Size & Position controls to set a map frame size with a width of **5 in** and height of **7 in**.

9. Move the Max Advertised Broadband Speed map frame and snap it to the guide. Attempt to top-align the frame with the other map frame. Do not worry about modifying the anchor settings manually.

As you noticed, smart guides appear as you move the map frame to help you align it with the other map frame. You will modify the different map extents next.

10. With the Max Advertised Broadband Speed map frame still selected, right-click and click Activate. (Alternatively, on the Layout tab, in the Map group, click Activate.)

In activated map mode, you can work with the map within the context of the page. Notice that the contextual controls change to the same commands as they do in a *map view*, and a Layout link appears in the Layout pane, directly above the ruler. Another set of navigation tools becomes available specifically for the layout.

11. On the Map tab, go to the Utah County bookmark.

Do not worry that your bookmarks are all grouped under "Community Anchor Institutes Bookmarks."

12. Click the arrow or the Layout link in the upper corner of the view to close the activated map mode.

13. Activate the Community Anchor Institutes map frame, go to the Provo bookmark, and close the activated mode when finished.

Both maps now show the areas of interest for your project.

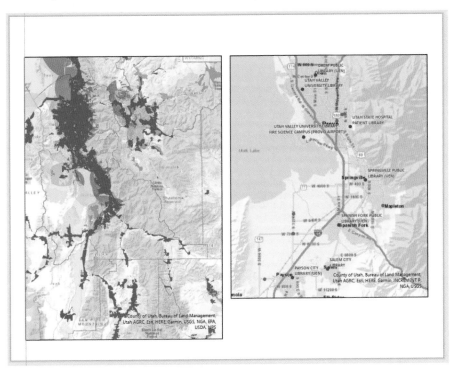

14. In the Contents pane, rename Map Frame to **Community Anchor Institutes**, and rename Map Frame 1 to **Max Advertised Broadband Speed**.

Keyboard shortcuts are useful and can help you save time when working with layouts. Generally, the layout view must be in focus to use the keyboard shortcuts. For more information, go to ArcGIS Pro Help > Layouts > "Keyboard Shortcuts for Working on the Layout."

Insert a legend element

A *legend* will help your map readers understand the meaning of the broadband speed symbols on the map as well as the status of public Wi-Fi availability at libraries in Utah County. A basic legend consists of a patch (symbol) and a label (explanatory text).

1. With the Max Advertised Broadband Speed map frame selected, click the Insert tab. In the Map Surrounds group, click Legend.
    The pointer changes to cross hairs. You will draw a rectangle to create the legend.

2. Drag a rectangle on the layout to draw the legend anywhere in the layout view. Make sure to create a reasonably sized rectangle as a small rectangle will not show any legend elements.

    A Legend item appears in the Contents pane. Because the Wireline_Broadband and Wireless_Broadband layers use the same symbology, it is not necessary to list the symbology for both layers in the legend.

3. With the legend selected, go to the Contents pane. For Legend, clear the check boxes for the first Wireline_Broadband item and the Counties item. Alternatively, you can also remove the layers.

    Next, you will adjust the properties of the legend.

    Some layout tools leave the pointer in a tool mode, such as when drawing a rectangle. To return to the select mode, on the Layout tab, click Select.

4. Right-click Wireless_Broadband and click Properties.

5. In the Wireless_Broadband Element pane, for Sizing, change Patch Width to **16 pt** and Patch Height to **10 pt**.

6. For Show, clear the Layer Name and Headings check boxes.

    These headings are not necessary because there is only one set of symbols in the legend. Layer and heading names are useful when multiple groups of symbols are represented.

7. Click Text Symbol to view additional options. On the General tab, for Appearance, change Font Name to **Arial** and Size to **9 pt**. Click Apply.

    Next, you will add a background for the legend so that it stands out when placed on the map frame.

8. Click the back button twice to return to the Legend options. From the four option tabs at the top, click the Display tab, and adjust the following properties:
    - For Border, change the symbol to **0 pt**.
    - For Background, click the Fill down arrow, and click Arctic White.
    - Click the Fill down arrow again and click Color Properties.
    - In the Color Editor, change the Transparency to **30%** and click OK.
    - To adjust the gap to improve the spacing around the patches and labels, increase the Background X and Y gaps to **0.1** inches.

9. Resize the legend, and position it near the upper-right corner of the Max Advertised Broadband Speed map frame (an appropriate size is 1.6 by 1.5 inches).

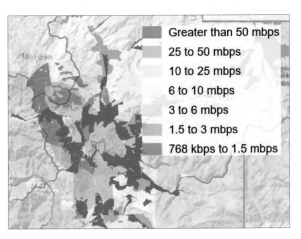

To nudge the legend into place, press and hold Ctrl, and use the arrow keys to move it.

ON YOUR OWN

Use the same steps to create a legend for the Community Anchor Institutes map. (Hint: Activate the Community Anchor Institutes frame first and edit the Heading text.) Next, insert the legend, adjust the group heading font, and display Legend Items. In the Contents pane, remove Counties from the display. Adjust the background and position the legend in the upper-right corner of the map frame.

Insert a scale bar

A *scale bar* is a dynamic element that provides an indication of the size of features and distances on the map. You will add a scale bar for each map frame.

1. With the Max Advertised Broadband Speed map frame selected, click the Insert tab. In the Map Surrounds group, click the Scale Bar down arrow. In the gallery, under Imperial, click Scale Line 3. Drag a rectangle on the layout to place the scale.

    A scale bar based on the Scale Line 3 style is placed on the page. A Scale Bar item appears in the Contents pane. Next, modify the scale bar to suit the map frame.

2. With the scale bar selected, click the Scale Bar tab on the ribbon.

3. In the Text Symbol group, change the font size to **8 pt**.

4. Click the Design tab on the ribbon.

5. In the Divisions group, change Divisions to **1**. In the Units group, make sure that Units is set to Miles.

6. Double-click the scale bar to open the Scale Bar pane, and adjust the following properties:
    *   Similar to the legend, create a white background with **30%** transparency.
    *   Resize the scale bar so that it displays to **20 miles**. To do so, zoom in substantially.
    *   Position the scale bar near the lower-left corner of the Max Advertised Broadband Speed map frame.

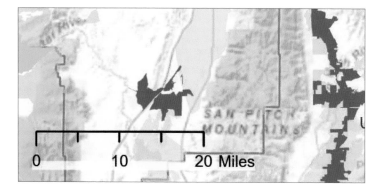

7. In the Contents pane, right-click the existing scale bar element and click Copy. Right-click Layout and click Paste.

A duplicate of the scale bar is added to the page. It sits on top of the existing scale bar. You will modify it to match the map scale of the other map frame.

8. Move the duplicated scale bar to the other map frame, and position it near the lower-left corner. On the Design tab on the ribbon, in the Map Frame group, choose Community Anchor Institutes Map frame from the list.

The scale bar changes to match the map scale of this map frame.

9. Resize the scale bar to display **four miles**, zooming in, if necessary.

Both scale bars are configured and placed according to the map scale of each map frame. Next, you'll insert more map elements to customize your layout.

Insert a north arrow

A *north arrow* is a dynamic element that indicates the orientation of the map.

1. On the Insert tab, click the North Arrow down arrow, and click ArcGIS North 2.

2. Click to place the north arrow. Position it below the Community Anchor Institutes map frame. Resize it so that it is clear and visually pleasing.

3. Double-click the north arrow. In the North Arrow Element pane, on the Options tab, click the north arrow symbol. Change the fill color to Gray **60%** and click Apply.

4. Click the north arrow on the map and align it just off the bottom of the map, to the left. If snapping is on, snap it to the invisible grid.

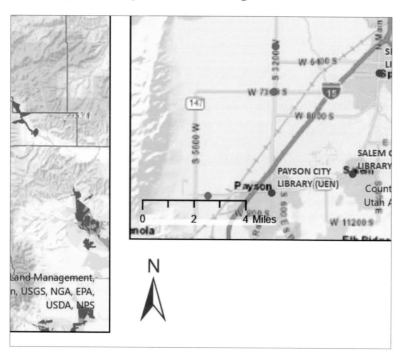

Insert dynamic text, a title, and rectangles

You will finish the layout with some descriptive text elements and decorative shapes.

1. On the Insert tab on the ribbon, in the Graphics and Text group, click the Dynamic Text down arrow and click Spatial Reference.

2. Draw a rectangle in the layout to place the dynamic text element. Right-click the dynamic text element and click Convert to Graphics.

   Although *dynamic text* is useful, in this case there is more information than required. You will remove some of the text so that it shows only the coordinate system information.

3. In the Text Element pane, keep only the PCS, GCS, and Projection tags and delete the others.

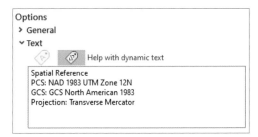

This displays current properties of the project. When a property is updated, the text automatically updates.

4. Position the spatial reference information element next to the north arrow.

5. On the Insert tab, in the Graphics and Text group, click Rectangle Text. Draw a rectangle anywhere in the layout to place the text box. In the text box in the pane, type **Maximum Advertised Download Speeds Utah County**. (Hint: Press Shift + Enter to make a line break and put "Utah County" on a separate line.) Click anywhere outside the text box to enter the text.

6. Adjust the following properties:
   - In the Element Text 1 pane, click Text Symbol.
   - On the General tab, under Appearance, change the font to **Calibri** and the font size to **18 pt**.
   - Expand Position, change Horizontal Alignment to Center, and apply the changes.

7. Center the text element above the left map frame.

> ### ON YOUR OWN
> Using the same steps, create a title for the Community Anchor Institutes map frame using the title **Public Wi-Fi Status at Libraries Utah County**. Add dynamic text elements that may be relevant (such as user, date, or project folder path).

8. On the Insert tab, in the Graphics and Text group, under Polygon, click Rectangle, and draw a rectangle over a map title to place it in the layout.

9. In the Element Rectangle pane, click the Symbol tab, and set the Color to 20% Gray and the Outline Color to No Color and click Apply.

The drawing order of layout elements works the same as for map layers. You want to make the title text visible over the gray rectangle.

10. In the Contents pane, move the Rectangle element below the Text elements.

11. Using the same process, add a 20% gray title background to the other map. Adjust and resize the rectangles as needed to match the accompanying figure.

12. With snapping turned on, snap both text elements and both gray rectangles to appropriate borders (the width of each map frame).

13. On the Layout tab, in the Show group, clear the Rulers and Guides check boxes. In the Review group, click Check Spelling.

14. Save the project.

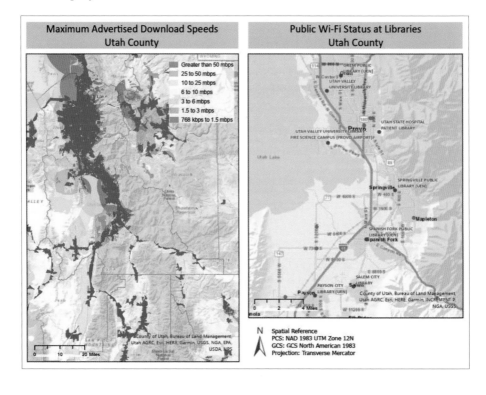

Your page layout should look good, with all the elements where they should be. Page layouts take time to create, and when done properly, they can communicate information effectively. The next step is to share your layout.

Exercise 9d: Share your project

Estimated time to complete: 20 minutes

After creating your layout, you can export it to a file or share it online. You can share it as a file several ways. A PDF file is one of the most popular industry-standard formats; it is editable in many graphics applications and retains map georeference information, labeling, and feature attribute data.

Exercise workflow

- Save the map as a geospatial PDF file.
- Save the project as a layout file.
- Share your project template to ArcGIS Online.

Export a layout to a PDF file

1. Make sure that the Layout view is the active view. On the Share tab, in the Output group, click Export Layout.

2. In the Export Layout pane, for File Type, click PDF. For Name, browse to your project folder, name the file **Utah County Broadband Map**, and click OK.

3. At the bottom of the Export Layout pane, under PDF Settings, verify that the Export Georeference Information check box is checked. For Layers and Attributes, click PDF Layers Only. Click Export.

   The export may take a few moments to finish.

4. When the export is completed, at the bottom of the Export Layout pane, click View Exported File.

   You can use Adobe Acrobat tools to find and mark location coordinates; measure distance, perimeter, and area; change coordinate system and measurement units; and copy location coordinates to the clipboard. Then you can use these tools to show locations in several web-mapping services.

OTHER EXPORT FORMATS

You can export to EMF, EPS, PDF, SVG, and SVGZ formats, which support a mixture of vector and raster data. You can also export to BMP, JPEG, PNG, TIFF, TGA, and GIF image formats, which are purely raster data. Most raster formats can include a world file that contains georeferenced information so that it can be used in ArcGIS Pro or other GIS applications. (This option is available only for map exports.) The TIFF format can retain georeferenced information internally—no external reference file is required—and is known as a GeoTIFF.

Save a layout file

A layout file includes the page, layout elements, and any maps referenced by map frames on the page. You can use a layout file to create templates or share existing layouts with others. However, data is not included in the layout file—only links to the data. Whomever you share the data with must have the data in the same folder location, or they can simply point to the data to repair the source.

1.   Make sure that the Layout view is the active view. On the Share tab, in the Save As group, click Layout File.

2.   Browse to your project folder, and name the file **Utah County Layout**.

     Layout files are saved using a .pagx extension. Whomever you shared your layout file with will see the same layout as the one that was saved, and the surrounding map elements and dynamic text will be updated accordingly. Layout files are often small and can be shared easily, including with your organization in ArcGIS Online.

Package and share a project template

Although page layout files are useful, they do not include data. To truly share aspects of your Utah broadband map project, you can create a project template, which can be shared directly in ArcGIS Pro and uploaded to your ArcGIS Online organization in only a few steps.

Your project template will include the two maps you worked on in exercises 9a through 9c, page layouts, and a connection to the database used. You can include attachments, such as image files or supporting documentation. Depending on how you share the template, another user from your organization can open the project directly in ArcGIS Pro.

> *Always consider how you want project templates to be shared. Do not share sensitive data or projects with everyone in ArcGIS Online. Often, it is safer to first share to your My Content folder and then manage which group or organization you want to share with using the sharing options in ArcGIS Online.*

> *You can also share as a project package that includes not only all maps and layer data but also the folder connections, toolboxes, geoprocessing history, and attachments. Project packages are an excellent way to share entire projects with colleagues, generate a snapshot of a project for archiving purposes, or share the project outside your organization.*

1. On the Share tab, in the Save As group, click Project Template.
   The Create Project Template pane appears.

2. For Start Creating, click Upload Template to Online Account. (Uploads require an internet connection. If you do not want to upload the project template, click Save Template to File.)

3. Change the name to **Utah Broadband Project**. (If you are saving it to disk, choose your project folder, and save it as **Broadband.aptx**.)

4. For Summary, type **A broadband project template for Utah counties**. In the Tags box, type **Broadband, Utah, Libraries** (comma separated so that individual tags are recognized). Press Enter.

   Under Share With, the organizations and groups you belong to are listed. For this exercise, you will not share the project template, so no further action is needed. Remember that you can change how your projects and content are shared in ArcGIS Online.

5. Click Analyze to check for errors. If there are any errors or warnings, go back to the Template tab and correct them.

6. Once validated, click Create to simultaneously create the project template and upload it to your ArcGIS Online account. Click Yes if you are asked to save the project.

ArcGIS Pro begins to package, compress, and upload the project template. Depending on the speed of your computer and internet connection speed, it may take a few minutes.

7. Exit ArcGIS Pro.

ON YOUR OWN

In a web browser, go to ArcGIS Online and sign in. Go to your My Content workspace to see if the Utah Broadband Project successfully uploaded. From there, share with other members of your organization or any groups to which you belong.

Summary

You have successfully created a layout that includes two maps of Utah County, one to show maximum advertised broadband speeds and one to show the status of available public Wi-Fi at libraries. From start to finish, you have created detailed symbology, including using a definition query to focus on specific attributes of the data and a label class to label specific libraries based on an unknown public Wi-Fi status. You also now have a foundational skill set for creating outputs that can be shared with other members of your organization. Although PDF and image exports are useful for distributing maps and information, your organization may also want to work with the data on its own, using project templates. The project template you shared to ArcGIS Online is just one step of a workflow. An organization can use many additional ArcGIS Online tools to manage data and allow users to work with content. You should now be comfortable in the ArcGIS Pro environment and ready to begin working with your own projects.

Glossary terms

layout
label
label class
map frame
map extent
map view
legend
scale bar
north arrow
dynamic text

Glossary

This glossary contains definitions for the ArcGIS terms used throughout the book and noted in the list at the end of each chapter. Many of these definitions can also be found in the Esri GIS Dictionary at https://support.esri.com/en/other-resources/gis-dictionary.

ArcGIS Pro–specific terminology can be found online in ArcGIS documentation at https://pro.arcgis.com/en/pro-app/get-started/get-started.htm.

address locator. A dataset in ArcGIS that stores the address attributes, associated indexes, and rules that define the process for translating nonspatial descriptions of places, such as street addresses, into spatial data that can be displayed as features on a map. An address locator contains a snapshot of the reference data used for geocoding and includes the parameters for standardizing addresses, searching for match locations, and creating output.

alias. An alternative name specified for fields, tables, files, or datasets that is more descriptive and user-friendly than the actual name.

altitude. The height above the horizon, measured in degrees, from which a light source illuminates a surface. Altitude is used when calculating a hillshade, or for controlling the position of a light source in a scene.

ArcGIS Pro. An application for creating and working with spatial data on your desktop. It provides tools to visualize, analyze, compile, and share your data, in both 2D and 3D environments.

aspect. The compass direction that a topographic slope faces, usually measured in degrees from north. The aspect can be generated from continuous elevation surfaces.

attribute. Nonspatial information about a geographic feature in a GIS, usually stored in a table and linked to the feature by a unique identifier. For example, attributes of a river might include its name, length, and sediment load at a gauging station. Attribute tables also store information about feature geometry, such as length and area.

attribute domain. In a geodatabase, a mechanism for enforcing data integrity. Attribute domains define what values are allowed in a field in a feature class or a nonspatial attribute table. If the features or nonspatial objects are grouped into subtypes, different attribute domains can be assigned to each of the subtypes.

attribute join. Appending the fields of one table to those of another through an attribute common to both tables. See also spatial join and relate.

attribute query. A request for records of features in a table based on their attribute values.

azimuth. In GIS software, the direction from which a light source illuminates a surface.

basemap. A map depicting background reference information, such as landforms, roads, landmarks, political boundaries, or satellite imagery, onto which other thematic information is placed.

buffer. A zone around a map feature measured in units of distance or time. A buffer is useful for proximity analysis.

cell. The smallest unit of information in raster data, usually square in shape. In a map or GIS dataset, each cell represents a portion of the earth, such as a square meter or square mile, and usually has an attribute value associated with it, such as soil type or vegetation class.

clip. A command that extracts features from one feature class that reside entirely within a boundary defined by features in another feature class.

command line. A programming environment in which the user enters commands by means of strings of text typed on a keyboard, as opposed to selecting commands from graphic prompts such as icons or dialog boxes.

continuous data. Data such as elevation or temperature that varies without discrete steps. Because computers store data discretely, continuous data is usually represented by TINs, rasters, or contour lines, so that any location has either a specified value or one that can be derived.

coordinate system. A reference framework consisting of a set of points, lines, or surfaces and a set of rules used to define the positions of points in space in either two or three dimensions. The Cartesian coordinate system and the geographic coordinate system used on the earth's surface are common examples of coordinate systems.

data type. The attribute of a variable, field, or column in a table that determines the kind of data it can store. Common data types include character, integer, decimal, single, double, and string.

defined interval classification. Also known as class interval. A set of categories for classification that divide the range of all values so that each piece of data is contained within a nonoverlapping category.

definition query. A request that examines feature or tabular attributes based on user-selected criteria and displays only those features or records that satisfy the criteria.

discrete data. Data that represents phenomena with distinct boundaries. Property lines and land-use areas are examples of discrete data.

dissolve. A geoprocessing command that removes boundaries between adjacent polygons that have the same value for a specified attribute.

dynamic text. Text placed on a layout that changes based on the current properties of an element, such as a username, the date of a project, the file path of the project, the metadata of a map on your page, and so on.

edge. A line between two points that forms a boundary.

endpoints. The points that define the length of a line segment.

equal interval classification. A data classification method that divides a set of attribute values into groups that contain an equal range of values.

extrusion. The process of stretching a flat 2D shape vertically to create a 3D object.

feature class. In ArcGIS, a collection of geographic features that have the same geometry type (such as point, line, or polygon), the same attributes, and the same spatial reference. Feature classes can be stored in a geodatabase, shapefile, or coverage, among other data formats. Feature classes allow homogeneous features to be grouped into a single unit for data storage purposes. For example, highways, primary roads, and secondary roads can be grouped into a line feature class named "roads." In a geodatabase, feature classes can also store annotation and dimensions.

feature dataset. A collection of feature classes stored together that share the same spatial reference; that is, they share a coordinate system, and their features fall within a common geographic area. Feature classes of different geometry types can be stored in a feature dataset.

feature template. A named set of tools that creates new features. Feature templates contain properties that you can configure and define how a new feature is created.

geocoding. A GIS operation for converting street addresses into spatial data that can be displayed as features on a map, usually by referencing address information from a street segment data layer.

geodatabase. A database or file structure used primarily to store, query, and manipulate spatial data. Geodatabases store geometry, a spatial reference system, attributes, and behavioral rules for data. Various types of geographic datasets can be collected within a geodatabase, including feature classes, attribute tables, raster datasets, network datasets, topologies, and others. Geodatabases can be stored in relational database management systems or in a system of files, such as a file geodatabase.

geographic coordinate system. A reference system that uses latitude and longitude to define the locations of points on the surface of a sphere or spheroid. A geographic coordinate system definition includes a datum, prime meridian, and angular unit.

geometric interval classification. A classification scheme in which the class breaks are based on class intervals that are in a geometric series. The geometric coefficient in this classifier can change once (to its inverse) to optimize class ranges.

geoprocessing. A GIS operation used to manipulate GIS data. A typical geoprocessing operation takes an input dataset, performs an operation on that dataset, and returns the result of the operation as an output dataset. Common geoprocessing operations include geographic feature overlay, feature selection and analysis, topology processing, raster processing, and data conversion. Geoprocessing helps define, manage, and analyze the information used to form decisions.

GIS. Acronym for geographic information system. An integrated collection of computer software and data used to view and manage information about geographic places, analyze spatial relationships, and model spatial processes. A GIS provides a framework for gathering and organizing spatial data and related information so that it can be displayed and analyzed.

graduated colors. A way to symbolize map features in which a range of colors indicates a progression of numeric values. For example, increases in population density may be represented by the increased saturation of a single color, or temperature differences may be shown by a sequence of colors from blue to red.

graduated symbols. A way to symbolize point or line features according to the value of the attribute they represent. For example, denser populations may be represented by larger dots or larger rivers by thicker lines.

hillshade. Shadows drawn on a map to simulate the effect of the sun's rays over the varied terrain of the land.

hot spot analysis. An analysis that attempts to identify statistically significant spatial clusters of high values (hot spots) and low values (cold spots).

input table. The existing attribute table for a target layer in a join operation. To append another table to an input table, the two tables must share at least one attribute field.

intersect. A geometric integration of spatial datasets that preserves features or portions of features that fall within areas that are common to all input datasets.

intersection. The point where two lines cross. In geocoding, most often a street crossing.

join table. The table in a join operation, often a nonspatial table, that is appended to the target layer's attribute table.

kernel density. Calculates a magnitude-per-unit area from point or polyline features using a kernel function to fit a smoothly tapered surface to each point or polyline.

label. In ArcGIS, descriptive text, usually based on one or more feature attributes. Labels are placed dynamically on or near features based on user-defined rules and in response to changes in the map display. Labels cannot be individually selected and modified by the user. Label placement rules and display properties (such as font size and color) are defined for an entire layer.

label class. Label classes can be used to restrict labels to certain features or to specify different label fields, symbols, scale ranges, label priorities, and sets of label placement options for different groups of labels.

layer. In ArcGIS, a reference to a data source, such as a shapefile, coverage, geodatabase feature class, or raster, that defines how the data should be symbolized on a map. Layers can also define additional properties, such as which features from the data source are included. Layers can be stored in a project (.aprx), or saved individually as a layer file (.lyrx).

layer file. A file with a .lyrx extension that stores the path to a source dataset and other layer properties, including symbology.

layer package. A file (.lpkx) that includes both the layer properties and the source data referenced by the layer. Layer packages contain symbolization, labeling, table properties, and the data. They can be created and viewed in ArcGIS Pro, ArcMap, ArcGlobe, and ArcScene, as well as viewed in ArcGIS Online.

layout. A collection of organized elements. In a layout, common elements, including the map, map labels, a title, legend, scale bar, north arrow, captions, and additional graphics, can be arranged.

legend. The description of the types of features included in a map, usually displayed in the map layout. Legends often use graphics of symbols or examples of features from the map with a written description of what each symbol or graphic represents.

location query. Also called a spatial query, it is a statement or logical expression that selects geographic features based on location or spatial relationship. For example, a spatial query might find what points are contained within a polygon or set of polygons, find features within a specified distance of a feature, or find features that are next to each other.

manual interval classification. The process of sorting or arranging entities into groups or categories; on a map, the process of representing members of a group by the same symbol, usually defined in a legend.

map. The visual display of geographic data on paper or a screen.

map algebra. A language that defines a syntax for combining map themes by applying mathematical operations and analytical functions to create new map themes. In a map algebra expression, the operators are a combination of mathematical, logical, or Boolean operators (+, >, AND, OR, tan, and so on) and spatial analysis functions (slope, shortest path, spline, and so on), and the operands are spatial data and numbers.

map extent. The limit of the geographic area shown on a map, usually defined by a rectangle. In a dynamic map display, the map extent can be changed by zooming and panning.

map frame. A container for maps in your layout, which can reference any map or scene in your project. Empty map frames can be used when creating templates. The map extent of a map frame is independent of any map view in the project. A dataset can be represented in one or more map frames. In map view, only one map frame is displayed at a time; in a layout, multiple map frames can be displayed at the same time.

map projection. A method by which the curved surface of the earth is portrayed on a flat surface. Map projection generally requires a systematic mathematical transformation of the earth's graticule of lines of longitude and latitude onto a plane. Some projections can be visualized as a transparent globe with a light bulb at its center (although not all projections emanate from the globe's center) casting lines of latitude and longitude onto a sheet of paper. Generally, the paper is either flat and placed tangent to the globe (a planar or azimuthal projection) or formed into a cone or cylinder and placed over the globe (a cylindrical or conical projection). Every map projection distorts distance, area, shape, or direction, or some combination thereof.

map view. An all-purpose view for exploring, displaying, and querying geographic data. This view does not show any map elements, such as titles, north arrow, or scale bars.

mask. A means of identifying areas to be included in analysis. Such a mask is often referred to as an analysis mask and may be either a raster layer or feature layer.

merge. A geoprocessing command that combines multiple input datasets into a single, new output dataset. A merge can combine point, line, or polygon feature classes or tables.

metadata. Information that describes the content, quality, condition, origin, and other characteristics of data or other pieces of information. Metadata for spatial data may describe and document its subject matter; how, when, where, and by whom the data was collected; availability and distribution information; its projection, scale, resolution, and accuracy; and its reliability regarding some standard. Metadata consists of properties and documentation. Properties are derived from the data source (for example, the coordinate system and projection of the data), and documentation is entered by a person (for example, keywords used to describe the data).

model. In geoprocessing in ArcGIS, one process or a sequence of connected processes in ModelBuilder.

ModelBuilder. The interface used to build and edit geoprocessing models in ArcGIS.

natural breaks (Jenks) classification. A method of manual data classification that seeks to partition data into classes based on natural groups in the data distribution. Natural breaks occur in the histogram at the low points of valleys. Breaks are assigned in the order of the size of the valleys, with the largest valley assigned the first natural break.

NoData. In raster data, the absence of a recorded value. NoData does not equate to a zero (0) value. Although the measure of an attribute in a cell may be zero, a NoData value indicates that no measurement has been taken for that cell.

north arrow. A map symbol that shows the direction of north on the map, which shows how the map is oriented.

null hypothesis. A statement that essentially outlines an expected outcome when there is no pattern, no relationship, and/or no systematic cause or process at work; any observed differences are the result of random chance alone. The null hypothesis for a spatial pattern is typically that the features are randomly distributed across the study area. Significance tests help determine whether the null hypothesis should be accepted or rejected.

on-the-fly projection. ArcGIS applies the projected coordinate system of the first layer added to the map to all subsequent layers, ensuring that the data draws in the map's coordinate system.

open data. Data that is licensed to permit people to use the data freely. Open data can be reused, redistributed, and shared—even commercially.

operator. The symbolic representation of a process or operation performed against one or more operands in an expression, such as + (plus, or addition) and > (greater than). When evaluated, operators return a value as their result. If multiple operators appear in an expression, they are evaluated in order of their operator precedence.

organization. An ArcGIS Online organization is shared workspace composed of users, items, and groups. Users can view, contribute, and share web layers, maps, and other content.

overlay. In geoprocessing, the geometric intersection of multiple datasets to combine, erase, modify, or update features in a new output dataset.

project. In ArcGIS Pro, a collection of related geographic datasets, maps, layouts, tools, settings, and resources. A project is stored as an item in ArcGIS Online or on disk as an .aprx file.

projected coordinate system. A reference system used to locate x, y, and z positions of point, line, and area features in two or three dimensions. A projected coordinate system is defined by a geographic coordinate system, a map projection, any parameters needed by the map projection, and a linear unit of measure.

Python. A free, cross-platform, open-source programming language that is embedded in ArcGIS products and used for geoprocessing scripting.

quantile classification. A data classification method that distributes a set of values into groups that contain an equal number of values.

query expression. A type of expression that evaluates to a Boolean (true or false) value and is typically used to select those rows in a table whose values cause the expression to evaluate to true. Query expressions are generally part of a SQL statement.

raster. A spatial data model that defines space as an array of equally sized cells arranged in rows and columns and composed of single or multiple bands. Each cell contains an attribute value and location coordinates. Unlike a vector structure, which stores coordinates explicitly, raster coordinates are contained in the ordering of the matrix. Groups of cells that share the same value represent the same type of geographic feature.

reclassify. The process of taking input cell values and replacing them with new output cell values. Reclassification is often used to simplify or change the interpretation of raster data by changing a single value to a new value or grouping ranges of values into single values—for example, assigning a value of one (1) to cells that have values of 1 to 50, two (2) to cells that range from 51 to 100, and so on.

relate. Establishing a connection between two tables using an attribute that is common to both.

scale bar. A map element used to graphically represent the scale of a map. A scale bar is typically a line marked like a ruler in units proportional to the map's scale.

script. A small program or sequence of instructions, or the act of writing such a program; in ArcGIS, scripts are written using Python in a programming environment known as ArcPy.

select. To choose from a number or group of features or records; to create a separate set or subset.

shapefile. A vector data storage format for storing the location, shape, and attributes of geographic features. A shapefile is stored in a set of related files and contains one feature class.

snapping. An automatic editing operation in which points or features within a specified distance (tolerance) of other points or features are moved to match or coincide exactly with each other's coordinates.

space-time cube. A set of points summarized into a data structure by aggregation into space-time bins that form a cube. Within each bin, the points are counted, and specified attributes are aggregated.

spatial join. A type of table join operation in which fields from one layer's attribute table are appended to another layer's attribute table based on the relative locations of the features in the two layers.

spheroid. When used to represent the earth, a three-dimensional shape obtained by rotating an ellipse about its minor axis, either with dimensions that approximate the earth or with a part that approximates the corresponding portion of the geoid.

standard deviation classification. A data classification method that finds the mean value, and then places class breaks above and below the mean at intervals of either 0.25, 0.5, or 1 standard deviation until all the data values are contained within the classes. Values that are beyond three standard deviations from the mean are aggregated into two classes—greater than three standard deviations above the mean and less than three standard deviations below the mean.

style. An organized collection of predefined colors, symbols, properties of symbols, and map elements. Styles promote standardization and consistency in mapping products.

symbology. The set of conventions, rules, or encoding systems that define how geographic features are represented with symbols on a map. A characteristic of a map feature may influence the size, color, and shape of the symbol used.

task. A set of preconfigured steps that guide you and others through a workflow or business process. A task can be used to implement a best-practice workflow, improve the efficiency of a workflow, or create a series of interactive tutorial steps. Tasks reside in task items, which are stored within an ArcGIS Pro project.

task item. A series of steps, grouped into tasks, that walk you through a GIS workflow. A task item, composed of multiple tasks, might capture an entire workflow or one piece of a complex solution.

temporal data. Data that includes a time element and represents a state in time. Time values are stored in a single attribute field and can be used to visualize features or phenomena at moments on the timeline.

vector. A coordinate-based data model that represents geographic features as points, lines, or polygons. Each point feature is represented as a single coordinate pair, and line and polygon features are represented as ordered lists of vertices. Attributes are associated with each vector feature, as opposed to a raster data model, which associates attributes with grid cells.

vertex (vertices). One of a set of ordered x,y coordinate pairs that defines the shape of a line or polygon feature.

Task index

Add a basemap. 55
Add a layout to the project. 306
Add a new field. 95
Add data and create a bookmark. 60
Add data to a project 71
Add operations to ModelBuilder 174
Animate using the Range slider. 255
Animate using the Time slider 252
Apply informative symbols .84
Apply symbol layer drawing. 296
Author a task 164
Calculate field values 97
Calculate summary statistics 100
Change 3D visualization styling 247
Clip features 220
Clip raster layers 263
Combine criteria rasters 282
Configure filter and clustering 25
Configure map pop-up windows. 23
Configure snapping options. 127
Configure the map symbology 15
Convert a line feature to a polygon feature. 261
Convert a model to a geoprocessing tool 179
Convert shapefiles to geodatabase feature classes 116
Create a 3D scene 62
Create a definition query. .290

Create a folder connection 39
Create a kernel density 233
Create a line feature 131
Create a space-time cube 238
Create a spatial join 224
Create an address locator 205
Create buffers 217
Define the data 170, 182
Derive a hillshade surface 271
Derive an aspect surface 268
Derive a slope surface 269
Display a new field 99
Enable time 251
Enter attribute data 137
Establish an attribute domain 123
Examine feature attributes 46
Examine infographics 102
Examine the contextual ribbon 44
Explore a public map 11
Explore the map 43
Export a layout to a PDF file 317
Export the selection to a new dataset 76
Fill out the tool parameters 176
Fine-tune the symbology 293
Geocode addresses 207
Import a map document 38
Import layer symbology 88
Insert a legend element 309
Insert a north arrow 313
Insert a scale bar 311
Insert dynamic text, a title, and rectangles 314
Insert map frames 307
Join a table 199
Join data tables 80
Label features 52
Label features using Maplex Label Engine 300

Map x,y points 120
Measure distances. 54
Merge and dissolve features. 220
Merge polygons 143
Merge rasters 264
Modify feature symbols 49
Modify lines and points 146
Modify map contents 39
Overlay additional data 93
Package and share a project template. 318
Package and share the map 57
Package the project 191
Proceed through preconfigured tasks. 158
Reclassify criteria rasters. 278
Relate tables 104
Rematch addresses 211
Run a command using Python 184
Run the Emerging Hot Spot Analysis tool. .242
Run the model 178
Run the Optimized Hot Spot Analysis tool .235
Save a layout file 318
Save a map 30
Select by attribute and location. 222
Select by attributes 231
Select features 46
Select features by attributes 74
Set up a project. 156
Spatially join data 107
Split polygons 139
Start a new project. 35, 59
Switch to a local view 245
Symbolize using graduated colors 201
Use a custom script tool 189
Use the Swipe function to compare layers .92
Visualize a space-time cube. 239
Visually compare analysis outputs. 275

Image and data source credits

Image credits

Chapter 1

Abu Dhabi City, map courtesy of Municipality of Abu Dhabi City.

Ogallala Aquifer, map courtesy of Center for Geospatial Technology, data from US Geological Survey's Nebraska Water Science Center.

Senior Shedding in Portland, map courtesy of Portland State University; data from Oregon Health Division, US Census Bureau, Portland Metro's Regional Land Information System, US Geological Survey digital elevation models.

Chapter 3

Atlas of Heart Disease and Stroke, map created using the Interactive Atlas of Heart Disease and Stroke, a website developed by the Centers for Disease Control and Prevention, Division for Heart Disease and Stroke Prevention, http://nccd.cdc.gov/DHDSPAtlas.

Chapter 9

Continuous data map, data courtesy of the Office of Geographic Information (MassGIS) and the Commonwealth of Massachusetts Information Technology Division.

Discrete data map, data courtesy of Lake Tahoe Data Clearinghouse, US Geological Survey.

Scheid Vineyards photo, courtesy of Heidi Scheid.

Data credits

3D

Basemap courtesy of Esri, HERE, DeLorme, TomTom, Intermap, increment P Corp., GEBCO, USGS, FAO, NPS, NRCAN, GeoBase, IGN, Kadaster NL, Ordnance Survey, Esri Japan, METI, Esri China (Hong Kong), swisstopo, MapmyIndia, ©OpenStreetMap contributors, and the GIS User Community.

\GTKAGPro\3D\buildings.shp, courtesy of Department of Information Technology and Telecommunications (DoITT), New York City Department of City Planning.

Broadband

\GTKAGPro\Broadband\UtahBroadband.gdb, courtesy of the State of Utah Broadband Project, managed by Utah Automated Geographic Reference. Original data source is the National Broadband Map.

CityMaintain

\GTKAGPro\CityMaintain\Data\FireHydrants.shp, data courtesy of City of Troutdale, Oregon, Public Works Department.

\GTKAGPro\CityMaintain\Data\ WaterPressureZones.shp, data provided by Oregon Metro, Regional Land Information System (RLIS).

\GTKAGPro\CityMaintain\Data\Valves.dbf, data courtesy of City of Troutdale, Oregon, Public Works Department.

\GTKAGPro\CityMaintain\Data\WaterLines.shp, data courtesy of City of Troutdale, Oregon, Public Works Department.

\GTKAGPro\CityMaintain\Data\Wells.shp, data courtesy of City of Troutdale, Oregon, Public Works Department.

CommunityHousing

\GTKAGPro\CommunityHousing.gdb\census_ tracts, courtesy of US Census Bureau.

\GTKAGPro\CommunityHousing.gdb\ city_boundary, courtesy of Los Angeles County Enterprise GIS.

\GTKAGPro\CommunityHousing.gdb\la_census_ tracts, courtesy of US Census Bureau.

\GTKAGPro\CommunityHousing.gdb\social_ services, courtesy of Los Angeles County Enterprise GIS.

\GTKAGPro\CommunityHousing.gdb\streets, courtesy of Los Angeles County Enterprise GIS.

\GTKAGPro\CommunityHousing.tbx\household_ income.csv, courtesy of US Census Bureau, 2008–12 American Community Survey.

\GTKAGPro\CommunityHousing.tbx\public_ housing_sites.csv, courtesy of Los Angeles County Enterprise GIS, Public Housing Sites, Housing Authority of the City of Los Angeles, Housing Authority of the County of Los Angeles.

Conflict

Service layer basemap courtesy of Esri, HERE, DeLorme, TomTom, Intermap, increment P Corp., GEBCO, USGS, FAO, NPS, NRCAN, GeoBase, IGN, Kadaster NL, Ordnance Survey, Esri Japan, METI, Esri China (Hong Kong), swisstopo, MapmyIndia, ©OpenStreetMap contributors, and the GIS User Community.

\GTKAGPro\Conflict\Conflict.gdb\ ACLED_2005_2018_Africa. The Armed Conflict Location and Event Data (ACLED) project is directed by Professor Clionadh Raleigh (University of Sussex). It is operated by senior research manager Caitriona Dowd (University of Sussex). Andrew Linke is a consultant on the project (University of Colorado), while the data collection involves several research analysts, including Charles Vannice, James Moody, Daniel Wigmore-Shepherd, Andrea Carboni, and Roudabeh Kishi.

CrimeIncidents

\GTKAGPro\CrimeIncidents\CrimeIncidents.gdb\ crime, data courtesy of City of Philadelphia Police Department.

HealthStudy

Service layer basemap courtesy of Esri, HERE, DeLorme, TomTom, Intermap, increment P Corp., GEBCO, USGS, FAO, NPS, NRCAN, GeoBase, IGN, Kadaster NL, Ordnance Survey, Esri Japan, METI, Esri China (Hong Kong), swisstopo, MapmyIndia, ©OpenStreetMap contributors, and the GIS User Community.

\GTKAGPro\HealthStudy\Data\IL_food_deserts. shp, census tracts: TomTom, US Census, Esri. Food desert attributes: Economic Research Service (ERS), US Department of Agriculture (USDA).

\GTKAGPro\HealthStudy\Data\IL_med_income. shp, counties data courtesy of ArcUSA, US Census, Esri. Median income attributes courtesy of US Census Bureau, American Community Survey.

\GTKAGPro\HealthStudy\Data\Obesity_ Prevalence.dbf, data courtesy of Centers for Disease Control and Prevention (CDC).

\GTKAGPro\HealthStudy\Data\us_cnty_enc.shp, data courtesy of Esri, derived from TomTom, US Census.

Vineyard

\GTKAGPro\RetailSiteStudy, Retail Site Prospects—a text file that contains a compiled list of commercial properties, created by the author.

Service layer basemap courtesy of Esri, HERE, DeLorme, TomTom, Intermap, increment P Corp., GEBCO, USGS, FAO, NPS, NRCAN, GeoBase, IGN, Kadaster NL, Ordnance Survey, Esri Japan, METI, Esri China (Hong Kong), swisstopo, MapmyIndia, ©OpenStreetMap contributors, and the GIS User Community.

Student-added service layer basemap courtesy of Esri, DigitalGlobe, GeoEye, Earthstar Geographics, CNES/Airbus DS, USDA, USGS, AEX, Getmapping, Aerogrid, IGN, IGP, swisstopo, and the GIS User Community.

\GTKAGPro\Vineyard\Vineyard.gdb\ned_1, NED data produced by US Geological Survey.

\GTKAGPro\Vineyard\Vineyard.gdb\ned_2, NED data produced by US Geological Survey.

\GTKAGPro\Vineyard\Vineyard.gdb\planting_ sites, courtesy of Scheid Vineyards Inc., Greg Gonzales.

\GTKAGPro\Vineyard\Vineyard.gdb\property_ boundary, courtesy of Scheid Vineyards Inc., Greg Gonzales.

\GTKAGPro\Vineyard\Vineyard.gdb\Soils_clip, soils data produced by the US Geological Survey.

\GTKAGPro\Vineyard\Vineyard.gdb\vineyard_ blocks, courtesy of Scheid Vineyards Inc., Greg Gonzales.

World

Student-created service layer basemap courtesy of Esri, DeLorme, GEBCO, NOAA NGDC, DeLorme, HERE, Geonames.org, and other contributors.

\GTKAGPro\World\World.gdb\Cities, from Data and Maps for ArcGIS (2014), courtesy of ArcWorld.

\GTKAGPro\World\World.gdb\Countries, The World Bank: Air Pollution in World Cities (PM10 Concentrations): David R. Wheeler, Uwe Deichmann, Kiran D. Pandey, Kirk E. Hamilton, Bart Ostro, and Katie Bolt; from Data and Maps for ArcGIS (2014), courtesy of ArcWorld, US Census International Division, CIA Factbook.

\GTKAGPro\World\World.gdb\Latlong, from Data and Maps for ArcGIS (2014), courtesy of Esri.

\GTKAGPro\World\World.gdb\Ocean, from Data and Maps for ArcGIS (2014), courtesy of Esri.

About Esri Press

Esri Press is an American book publisher and part of Esri, the global leader in geographic information system (GIS) software, location intelligence, and mapping. Since 1969, Esri has supported customers with geographic science and geospatial analytics, what we call The Science of Where®. We take a geographic approach to problem-solving, brought to life by modern GIS technology, and are committed to using science and technology to build a sustainable world.

At Esri Press, our mission is to inform, inspire, and teach professionals, students, educators, and the public about GIS by developing print and digital publications. Our goal is to increase the adoption of ArcGIS and to support the vision and brand of Esri. We strive to be the leader in publishing great GIS books, and we are dedicated to improving the work and lives of our global community of users, authors, and colleagues.

Acquisitions
Stacy Krieg
Claudia Naber
Alycia Tornetta
Craig Carpenter
Jenefer Shute

Editorial
Carolyn Schatz
Mark Henry
David Oberman

Production
Monica McGregor
Victoria Roberts

Sales & Marketing
Eric Kettunen
Sasha Gallardo
Beth Bauler

Contributors
Christian Harder
Matt Artz

Business
Catherine Ortiz
Jon Carter
Jason Childs

For information on Esri Press books and resources, visit our website at
esri.com/en-us/esri-press.

Related titles

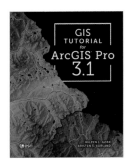

GIS Tutorial for ArcGIS Pro 3.1

Wilpen L. Gorr
and Kristen S. Kurland
9781589487390

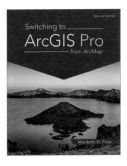

**Switching to ArcGIS Pro from
ArcMap,** *second edition*

Maribeth H. Price
9781589487314

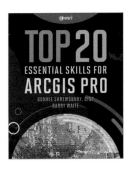

**Top 20 Essential Skills
for ArcGIS Pro**

Bonnie Shrewsbury
and Barry Waite
9781589487505

Getting to Know Web GIS,
fifth edition

Pinde Fu
9781589487277

For information on Esri Press books, e-books, and resources, visit our **website** at
esripress.com.